# ENERGY METHODS OF STRUCTURAL ANALYSIS

*By the same author*

ESSENTIAL SOLID MECHANICS
Theory, Worked Examples and Problems

# ENERGY METHODS OF STRUCTURAL ANALYSIS
Theory, worked examples and problems

B. W. Young

*Professor of Structural Engineering*
*Bayero University, Kano, Nigeria*

© B. W. Young 1981

All rights reserved. No part of this publication may be reproduced or transmitted, in any form or by any means, without permission.

*First published 1981 by*
THE MACMILLAN PRESS LTD
*London and Basingstoke*
*Companies and representatives throughout the world*

*Printed in Hong Kong*

ISBN 0 333 27776 7 pbk

The paperback edition of this book is sold subject to the condition that it shall not, by way of trade or otherwise, be lent, resold, hired out, or otherwise circulated without the publisher's prior consent in any form of binding or cover other than that in which it is published and without a similar condition including this condition being imposed on the subsequent purchaser.

# CONTENTS

*Preface* vii

1. PRELIMINARIES 1

    1.1 Definitions of Strain Energy and Complementary Energy 1

    1.2 The Basic Energy Theorems 2

    1.3 Potential Energy 5

    1.4 Stationary Complementary Energy 7

    1.5 Auxiliary Energy Theorems 9

2. FORCE AND DEFORMATION ANALYSIS OF PIN-JOINTED FRAMES 11

    2.1 Energy Due to Axial Forces 11

    2.2 Conversion of Statically Indeterminate Systems 12

    2.3 Choice of Method of Analysis 14

    2.4 The Compatibility Method 14

    2.5 The Equilibrium Method 37

    2.6 Deflexions in Pin-jointed Frames 50

    2.7 Design Example 60

    2.8 Problems 66

3. FORCE AND DEFORMATION ANALYSIS OF BEAMS, CURVED MEMBERS AND RIGID-JOINTED FRAMES 80

    3.1 Complementary Energy Due to Bending 80

    3.2 Straight Beams 81

    3.3 Curved Members 86

    3.4 Rigid-jointed Plane Frames 105

    3.5 Design Example 114

    3.6 Problems 120

| | | |
|---|---|---|
| 4. | POTENTIAL ENERGY METHODS | 128 |
| | 4.1 Conditions for Equilibrium : A Case Study | 128 |
| | 4.2 Structural Systems with a Limited Number of Degrees of Freedom | 133 |
| | 4.3 Approximate Solutions: The Rayleigh-Ritz Method | 134 |
| | 4.4 Design Example | 154 |
| | 4.5 Problems | 161 |
| *Index* | | 163 |

# PREFACE

A wide variety of different methods of structural analysis exist although many of them are designed for the solution of particular types of problem. Two procedures, however, are generally applicable; these are the method of virtual work and energy methods. In essence, the two methods are equivalent since, although the arguments used in establishing the governing equations differ, the equations themselves are identical.

In the author's experience, students studying virtual work find some difficulty in coming to terms with the idea of hypothetical forces or deformations whereas they quickly grasp the more obvious physical interpretations of the energy approach. One additional advantage that energy methods have over virtual work is their application to the approximate solution of complex problems for which exact solutions may not exist.

Energy methods have not received the consideration they deserve in the literature and it is for this reason that this book has been written. The few acceptable textbooks which treat the subject are at a level which is not easily accessible to undergraduates. It is a matter of regret that most undergraduate textbooks which do deal with energy methods generally reveal a hazy understanding of the principles and fail to take full advantage of their potential.

Energy methods have applications in almost every branch of structural analysis, as well as in many other fields. It is therefore essential that the undergraduate engineer should be fully conversant with their use. At the same time, it must be remembered that there are certain alternative specialised methods of analysis which might be quicker or easier for the solution of particular problems; it is the reader's responsibility to ensure that he does not put all his eggs in one basket.

This book is intended for second and third-year undergraduates in university or polytechnic degree courses. It should also prove useful as a source of reference to designers in practice. The format follows that of the author's previous book in this series, *Essential Solid Mechanics*; the theory, in concise form, is followed by a number of worked examples chosen to illustrate all the principles involved. Each chapter (except the first) ends with a selection of problems with answers which the reader may use for practice. The examples and problems are typical of those set in examinations at the end of the second or third years of a degree course. In some cases the precise origin of the questions is unknown but general acknowledgement is given here. The author alone is responsible for the solutions and answers.

One feature which will be of particular interest to both final-year undergraduates and practising designers is the design example which completes each of the three main chapters. Here an attempt has been made to create real-life design problems which may be solved by the application of energy methods. These examples must not be confused with the examination type of question.

B. W. YOUNG

ACKNOWLEDGEMENTS

For the proofs of Castigliano's first theorem (part I) and the first theorem of complementary energy, acknowledgement is made to Stephen Timoshenko and James Gere, on whose presentation (in *Mechanics of Materials*, Van Nostrand Reinhold, 1972) it would be impossible to improve.

# 1 PRELIMINARIES

Consideration of the energy stored in a structural system and the variation of this energy with force or deformation provides a powerful method of structural analysis.

A group of energy theorems are derived which apply to the general case of a structure made up of members having a non-linear (or linear) elastic stress-strain relationship. A further group of auxiliary theorems are then obtained for linear elastic behaviour as a special case.

To help with understanding the derivation and application of energy theorems, attention will be confined initially to pin-jointed plane frames which may be statically determinate or indeterminate. Later, applications of the theorems to a wider range of structural systems will be investigated.

## 1.1 DEFINITIONS OF STRAIN ENERGY AND COMPLEMENTARY ENERGY

The axial force-deformation curve for a non-linear elastic bar which is the qth member of a pin-jointed plane frame consisting of m members is shown in figure 1.1. The datum for the measurement of deformation is taken to be the initial, unstressed length of bar L.

The area $u_q$ under the force-deformation curve represents the energy stored in the member by the action of the force as it increases from zero to its final value, F. If the member is ideally elastic (exhibiting no hysteresis) all this energy is recoverable when the member is released from the structure.

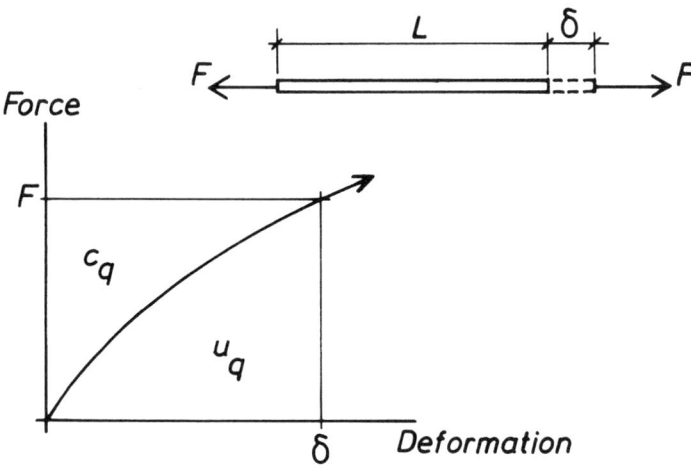

Figure 1.1

The area $c_q$ to the left of the curve has no direct physical meaning, but it can be seen that the total energy, $(u_q + c_q)$ represents the work done on the member by a constant force F acting through the deformation $\delta$. Thus

$$u_q + c_q = F\delta$$

The areas $u_q$ and $c_q$ have the dimensions of work or energy; $u_q$ is defined as the strain energy of the member and $c_q$ as the complementary energy.

## 1.2 THE BASIC ENERGY THEOREMS

We now examine a typical frame of which the bar of figure 1.1 might be a member. This frame is shown in figure 1.2 and is subjected to a set of external forces represented by loads $P_1, P_2, \ldots, P_n$ at the joints (only the first two loads and the last are shown in the figure for reasons of clarity). The displacement at the joints caused by the loads alone are $\Delta_1, \Delta_2, \ldots, \Delta_n$. These displacements are measured in the line of action of the corresponding load.

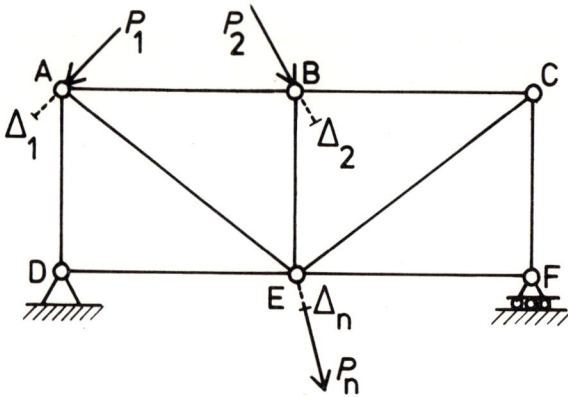

Figure 1.2

Figure 1.3 shows the relationship between one of the loads $P_j$ and its corresponding deflexion, $\Delta_j$ at a particular joint. To maintain generality the load-deflexion relationship is assumed to be non-linear elastic. The datum from which joint displacements are measured is taken to be the joint position prior to the application of any of the loads.

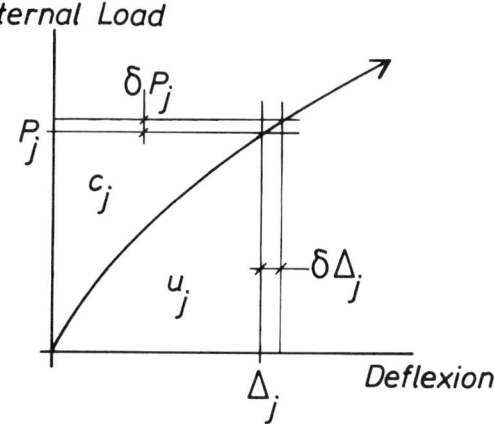

Figure 1.3

The areas $u_j$ and $c_j$ are defined as before and represent the contribution made by the load, $P_j$ and the corresponding deflexion, $\Delta_j$ to the total energy in the structure. The total work done by the loads, which is stored as an amount of strain energy $U_L$, is therefore given by

$$U_L = \sum_{j=1}^{n} u_j \qquad \text{(i)}$$

Similarly the total complementary energy $C_L$, stored by virtue of the action of the loads is given by

$$C_L = \sum_{j=1}^{n} c_j \qquad \text{(ii)}$$

where the summations extend over all the loaded joints.

The total strain energy, $U_m$ and the complementary energy, $C_m$ stored in the members can be expressed as

$$U_m = \sum_{q=1}^{m} u_q \qquad \text{(iii)}$$

and $$C_m = \sum_{q=1}^{m} c_q \qquad \text{(iv)}$$

where the summations extend over all m members in the frame. By consideration of the principle of conservation of energy it follows that

$$U_L = U_m = U \qquad \text{(v)}$$

and, since $U_L + C_L = U_m + C_m = \sum_{j=1}^{n} P_j \Delta_j$, we have

$$C_L = C_m = C \qquad \text{(vi)}$$

Imagine now that all save one of the joints in the frame of figure 1.2 are held in rigid clamps. If the load $P_j$ acting at the free joint is increased by a small amount $\delta P_j$ there will be a corresponding small increase, $\delta \Delta_j$ in the joint deflexion. The work done in producing this additional displacement is equal to an increase $\delta U$ in the total strain energy of the system. Thus from figure 1.3 and ignoring the product of the small quantities $\delta P_j$ and $\delta \Delta_j$, we have

$$\delta U = P_j \delta \Delta_j \qquad \text{(vii)}$$

Noting that the rate of change of strain energy in the system with respect to a particular deflexion $\Delta_j$ is given by the partial derivative $\partial U/\partial \Delta_j$, where U is a function of all the deflexions, we have the alternative expression for the change in strain energy given by

$$\delta U = \frac{\partial U}{\partial \Delta_j} \delta \Delta_j \qquad \text{(viii)}$$

Since the right-hand sides of equations (vi) and (viii) are identical, we have

$$\frac{\partial U}{\partial \Delta_j} = P_j \quad (j = 1, \ldots n) \qquad (1.1)$$

This equation is a statement of Castigliano's first theorem (part 1). Carlo Alberto Castigliano was an Italian engineer who derived the result in a book published in 1879.

Similarly, if all the loads but one are kept constant, the complementary work done during the resultant displacement is equal to an increase, $\delta C$ in the total complementary strain energy of the system. Again if the product of the small quantities is neglected, we have, from figure 1.3

$$\delta C = \delta P_j \Delta_j \qquad (ix)$$

The rate of change of complementary strain energy in the system with respect to a particular load $P_j$ is given by the partial derivative $\partial C/\partial P_j$, where C is a function of all the loads, hence

$$\delta C = \frac{\partial C}{\partial P_j} \delta P_j \qquad (x)$$

Since the right-hand sides of equations (ix) and (x) are identical, we have

$$\frac{\partial C}{\partial P_j} = \Delta_j \quad (j = 1, \ldots n) \qquad (1.2)$$

Equation 1.2 is a statement of the first theorem of complementary energy, usually attributed to Friedrich Engesser, a German engineer who derived the theorem in 1889. In fact, an Italian railway engineer, Francesco Crotti, had already obtained this result ten years earlier.

Equations 1.1 and 1.2 are perfectly general and apply to any elastic structural system. Castigliano's first theorem (part I), (equation 1.1) gives equations for the loads and requires the strain energy to be written in terms of the deformations. Its application is thus an example of the equilibrium approach to structural analysis. The first theorem of complementary energy (equation 1.2) on the other hand provides equations for the deformations and requires the energy to be expressed as a function of the loads. The application of this theorem is an example of the compatibility approach.

1.3 POTENTIAL ENERGY

To understand, in physical terms, the meaning of potential energy, it is useful to look again at figure 1.2 which represents a typical elastic structure deformed by a set of loads, $P_j (j = 1, \ldots n)$. The deformation of the structure is compatible with the displacements, $\Delta_j (j = 1, \ldots n)$ of the loads. The total work done by the loads as they increase from zero to their full value is stored in the structure as an amount of strain energy, U.

Suppose that, in imagination, we now apply an appropriate set of external forces (denoted by $\overline{P}$) to the loaded structure in order to return it to the original undeformed configuration. At the end of this process (during which the loads $P_j$ remain at their full value) the set of forces $\overline{P}$ are in equilibrium with the original loads $P_j$ and the structure itself is unstressed. It is important that this operation be carried out in a slow and controlled manner in order to eliminate inertia effects.

The potential energy of a deformed structure is defined as the total work done by the external force system, $\overline{P}$. This is not simply equal to the work done in moving the loads $P_j$ through their corresponding displacements $\Delta_j$; for an ideal system (no hysteresis), an amount of strain energy U is released in the form of useful work which does not have to be supplied by the force system $\overline{P}$.

Using the normally accepted sign convention, the work done *against* the loads $P_j$ (negative) is given by

$$(Wd)_1 = - \sum_{j=1}^{n} P_j \Delta_j \tag{i}$$

while the amount of work *given up* by the structure (positive) is

$$(Wd)_2 = U \tag{ii}$$

Hence, from the definition above, the potential energy of the system is given by

$$V = (Wd)_2 + (Wd)_1 = U - \sum_{j=1}^{n} P_j \Delta_j \tag{1.3}$$

If the strain energy is expressed in terms of characteristic displacements such as $\Delta_j$ we have, from equation 1.3

$$\frac{\partial V}{\partial \Delta_j} = \frac{\partial U}{\partial \Delta_j} - P_j \tag{iii}$$

but from Castigliano's first theorem (part I) (equation 1.1), for a structural system in equilibrium

$$P_j = \frac{\partial U}{\partial \Delta_j} \tag{iv}$$

thus from equations (iii) and (iv)

$$\frac{\partial V}{\partial \Delta_j} = 0 \tag{1.4}$$

This result is an expression of the principle of stationary potential energy which states that a structural system is an equilibrium if the displacements are such that the potential energy assumes a stationary value. The idea of a stationary value of the potential energy was first used by the German physicist, Gustav Kirchhoff in a paper published in 1850.

For stable equilibrium, V is a minimum, while for unstable

equilibrium V has a maximum value. For further discussion of stability, see chapter 4.

The principle of stationary potential energy applies to any elastic system (linear and non-linear) since it was derived solely by reference to Castigliano's first theorem (part I).

1.4 STATIONARY COMPLEMENTARY ENERGY

Consideration of complementary energy is particularly useful in determining forces in the redundant members of statically indeterminate systems.

Figure 1.4a shows a panel, ABCD which is part of a frame subjected to external loads.

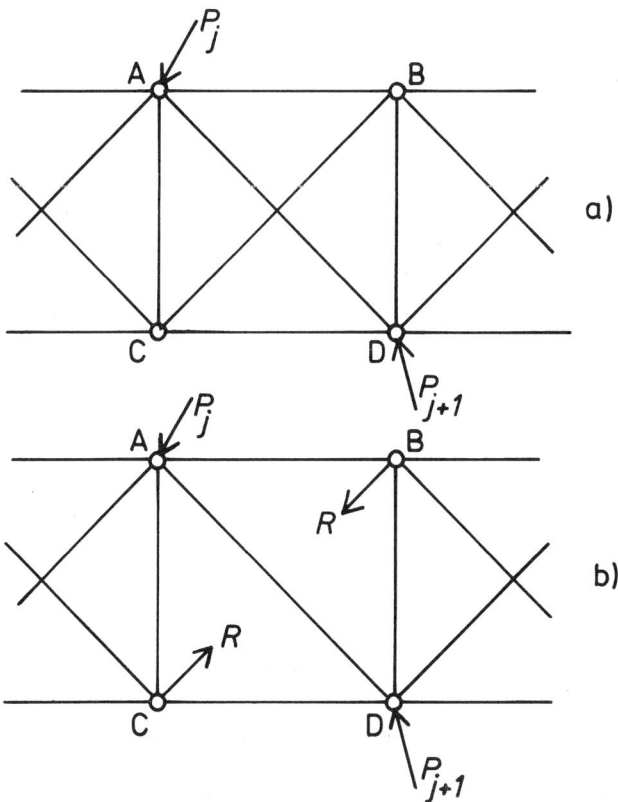

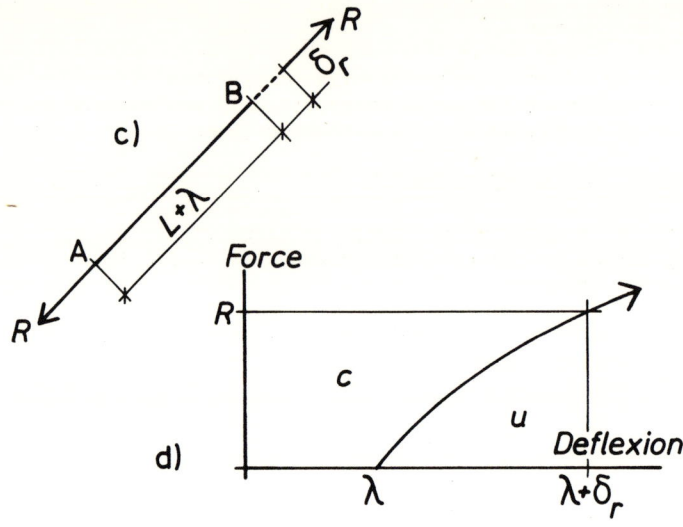

Figure 1.4

Suppose BC is taken to be the redundant member in the panel and let the unknown tensile force in BC be R. This member may therefore be replaced by a pair of equal and opposite forces R acting at B and C as shown in figure 1.4b.

If $C_1$ is the complementary energy of the whole frame, excluding member BC, we have from equation 1.2 that

$$\frac{\partial C_1}{\partial R} = \Delta_r \qquad\qquad (i)$$

where $\Delta_r$ is the amount B and C approach each other due to the action of all the forces on the frame. If L is the original distance between B and C in the unloaded frame, then the distance between B and C in the loaded frame becomes $(L - \Delta_r)$.

Now consider member BC, shown in figure 1.4c. The nominal length of BC is L, but due to a manufacturing error or a deformation brought about by a local temperature change, it has an actual length $L + \lambda$. Figure 1.4d shows the force-deformation diagram for member BC in which the nominal length L is taken as the datum. If the complementary energy in BC is c, from equation 1.2 we have

$$\frac{\partial c}{\partial R} = \lambda + \delta_r \qquad\qquad (ii)$$

8

where $\delta_r$ is the elastic deformation of BC due to the force R. The total length of BC is thus $L + \lambda + \delta_r$.

The loaded member BC must fit into the loaded frame, thus the two lengths determined above are equal and we have $L - \Delta_r = L + \lambda + \delta_r$, or

$$\Delta_r + \delta_r + \lambda = 0 \qquad \text{(iii)}$$

From equations (i), (ii) and (iii) it follows that

$$\frac{\partial C_1}{\partial R} + \frac{\partial c}{\partial R} = 0$$

But $(C_1 + c)$ is the total complementary energy, C, in the frame, thus for the general case where there are k redundancies

$$\frac{\partial C}{\partial R_i} = 0 \quad (i = 1, 2, \ldots, k) \qquad (1.5)$$

Equation 1.5 is an expression of the second theorem of complementary energy which states that the complementary energy of an initially unstressed, statically indeterminate structural system has a stationary value if the redundant forces are such as to ensure compatibility.

The presence of an initial lack of fit, $\lambda$, will produce forces in the members of a statically indeterminate structure even when no external forces act. This effect is known as 'self-straining'. By definition, no self-straining can occur in a statically determinate structure.

## 1.5 AUXILIARY ENERGY THEOREMS

Three further theorems involving strain energy may be derived for linearly elastic systems, for then (referring to figures 1.1 and 1.3), we have

$$c_q = u_q \text{ and } c_j = u_j$$

and from equations (i) to (vi) of section 1.2, equation 1.2 becomes

$$\frac{\partial U}{\partial P_j} = \Delta_j \quad (j = 1, 2, \ldots, n) \qquad (1.6)$$

This result is a statement of Castigliano's first theorem (part II).

An equation similar to 1.5 can be derived in terms of strain

9

energy. In this case, however, it must be remembered that no strain energy is stored until the elastic deformation $\delta_r$ occurs (see figure 1.4d). Thus the strain energy equivalent of equation (ii) in section 1.4 is

$$\frac{\partial U}{\partial R} = \delta_r$$

and the resulting general equation is

$$\frac{\partial U}{\partial R_i} = -\lambda_i \quad (i = 1, 2, \ldots, k) \tag{1.7}$$

which is sometimes referred to as Castigliano's second theorem.

Finally for no self-straining ($\lambda_i = 0$) in a linear elastic, statically indeterminate system, we have from equation 1.7 that

$$\frac{\partial U}{\partial R_i} = 0, \quad (i = 1, 2, \ldots, k) \tag{1.8}$$

This equation expresses the principle of stationary strain energy. It is sometimes referred to as the principle of least work.

Since the auxiliary theorems are not applicable in general, their use is not encouraged.

# 2 FORCE AND DEFORMATION ANALYSIS OF PIN-JOINTED FRAMES

The force analysis of a statically determinate pin-jointed frame is relatively simple. Sufficient equations for the unknown member forces may be obtained merely by satisfying the requirements of equilibrium. The treatment of statically indeterminate frames, on the other hand, is more difficult since more equations are required than are available from satisfaction of the equilibrium conditions.

One approach to the force analysis of statically indeterminate structural systems is the compatibility method, in which the first step is to derive equilibrium equations in terms of the unknown forces; the additional equations required as a result of the statical indeterminacy of the system are then obtained by satisfying the requirements of compatibility using the first or second theorems of complementary energy (equations 1.2 and 1.5).

An alternative approach is the equilibrium method in which the requirements of compatibility in the structure are first established. Extra equations are then obtained by satisfying the conditions of equilibrium using Castigliano's first theorem (part I) (equation 1.1).

Once the force analysis of a structure has been completed, it is a relatively straightforward matter to determine deflexions by application of the first theorem of complementary energy (equation 1.2).

The pin-jointed frame is, of course, an idealisation. In practice, joints will possess some degree of rigidity giving rise to bending moments having maximum values (for a given loading) in the case of perfectly rigid joints. These moments, however, are of secondary importance since the internal forces in the structure will arrange themselves in such a way that the strain energy is a minimum. For a given external force system, the least strain energy is stored if the internal forces cause axial rather than bending deformations (assuming gross deformations do not occur as in a collapsing strut).

## 2.1 ENERGY DUE TO AXIAL FORCES

Since this chapter is concerned solely with axial forces in members, it is useful to start by deriving appropriate expressions for strain energy and complementary energy.

Figure 2.1a shows a bar of cross-sectional area A and length L. The bar is subjected to an axial force F which produces a deformation $\Delta$. It will be assumed that the resulting stress is uniform and that deformations are small. Figure 2.1b shows the force-deformation curve for the bar. As we have already seen in chapter 1, the area under the curve represents the strain energy, U, stored in the bar, while the area above the curve represents the complementary energy, C, thus

$$U = \int F \, d\Delta$$

but $F = \sigma A$ and $\Delta = \varepsilon L$ where $\sigma$ and $\varepsilon$ are respectively the stress and strain in the bar. Hence

$$U = AL \int \sigma \, d\varepsilon$$

or
$$u = \int \sigma \, d\varepsilon \tag{2.1}$$

where u is the strain energy per unit volume.

Similarly, the complementary energy per unit volume, c, is given by

$$c = \int \varepsilon \, d\sigma \tag{2.2}$$

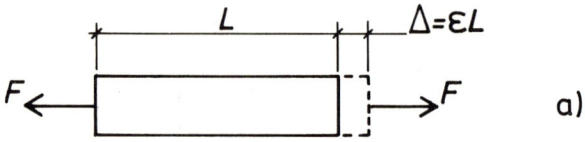

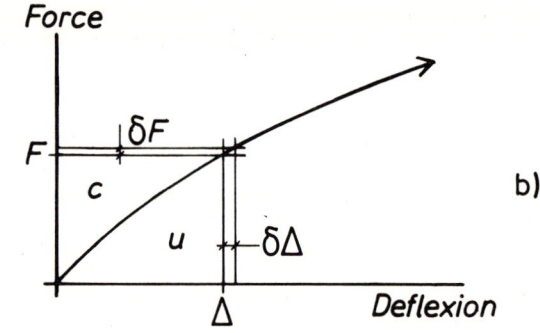

Figure 2.1

## 2.2 CONVERSION OF STATICALLY INDERTERMINATE SYSTEMS

The first step in the analysis of a statically indeterminate structure is to convert it into a number of statically determinate systems for which the force analysis is straightforward. The total forces in the original structure can then be obtained by application of the principle of superposition which, in the application below, is valid for both linear and non-linear elastic systems.

As an illustration, consider the redundant pin-jointed plane frame shown in figure 2.2.

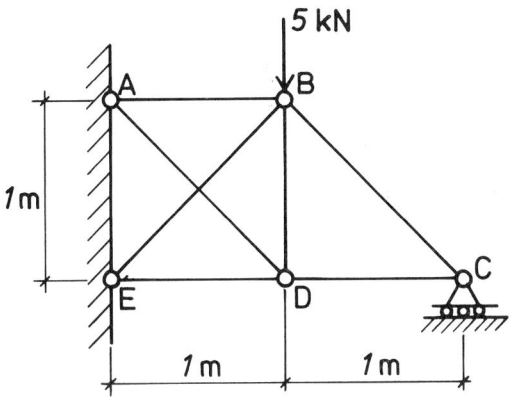

Figure 2.2

There are seven members (m), five joints (j) and five independent reactions (r). Thus m + r - 2j = 2 is the number of redundant forces in the frame (see Essential Solid Mechanics). It is convenient, though not obligatory, to take these redundant forces as the vertical reaction R at C and the force S in member AD.

A statically indeterminate system with N redundants may be converted into one statically determinate system carrying the external load with all the redundant forces set to zero and N statically determinate systems each subjected to only one of the redundant forces, all the other redundant forces being zero.

To simplify the calculation, it is usual to consider a unit value of the redundant force.

The statically determinate components of the frame in figure 2.2 are shown in figure 2.3.

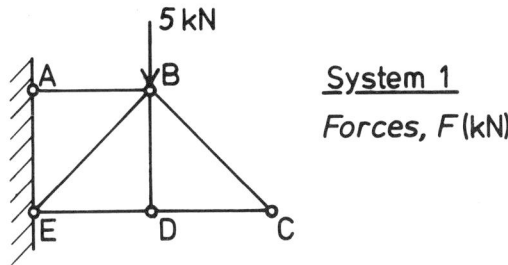

System 1
Forces, F (kN)

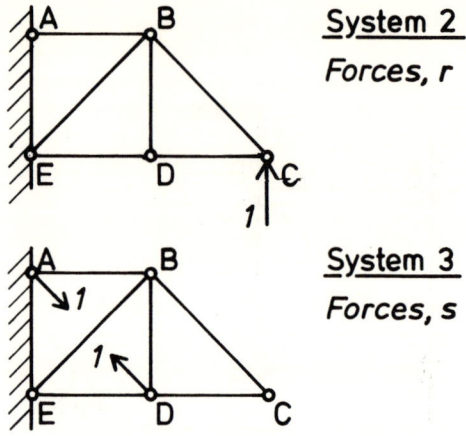

Figure 2.3

The total force, $F_T$, in any member of the original frame is obtained by superposition, thus

$$F_T = F + rR + sS$$

where r and s are the forces in the structure induced by unit values of the redundancies R and S respectively. Further discussion of this problem appears later in example 2.3.

## 2.3 CHOICE OF METHOD OF ANALYSIS

In the theoretical introduction (chapter 1), two fundamental methods for the analysis of structures were discussed. These were referred to as the equilibrium approach (Castigliano's first theorem (part I) and the principle of stationary potential energy) and the compatibility approach (first and second theorems of complementary energy). Either method may be used for a particular problem but as a general rule the compatibility approach is more efficient than the equilibrium method if the number of degrees of freedom of the structure is greater than the number of redundancies and vice versa.

## 2.4 THE COMPATIBILITY METHOD

Most redundant plane frames have more degrees of freedom than the number of redundancies, thus they are normally more amenable to the compatibility approach. We will consider here a number of examples of the use of this method starting with a simple problem in which the number of redundancies is equal to the number of degrees of freedom. In section 2.5, this same problem will be solved using the equilibrium approach to allow the reader to compare the two methods.

*Example 2.1*

Four linearly elastic rods made of the same material and having the same cross-sectional area are pin-jointed together at O. Their far ends are pinned to rigid supports at A, B, C and D as shown in figure 2.4. All the rods have the same length, L. If horizontal and vertical forces of 10 kN and 5 kN respectively, are applied to the joint O, determine the forces in the rods.

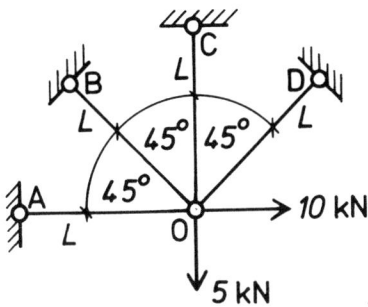

Figure 2.4

The frame has two redundant members which we will assume to be members OB and OD, thus we must consider the three statically determinate systems shown in figure 2.5.

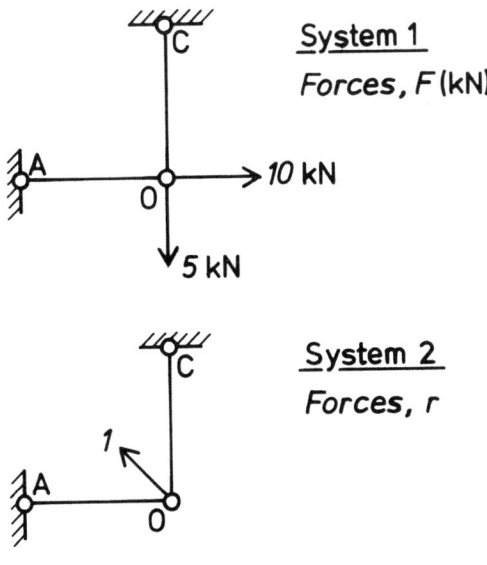

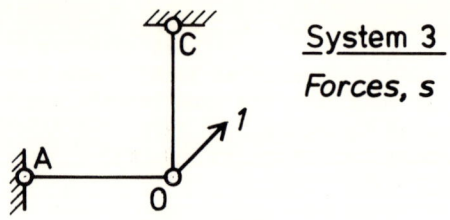

Figure 2.5

Let the force in OB be R and in OD be S. Then the total force, $F_T$, in a member of the original frame is given by

$$F_T = F + rR + sS \qquad \text{(i)}$$

The complementary energy for the whole frame is obtained from equation 2.2 as

$$C = \sum_1^4 \frac{F_T^2 L}{2AE} \qquad \text{(ii)}$$

thus from the second theorem of complementary energy (equation 1.5) we have

$$\frac{\partial C}{\partial R} = \sum_1^4 \frac{F_T L}{AE} \frac{\partial F_T}{\partial R} = 0$$

and

$$\frac{\partial C}{\partial S} = \sum_1^4 \frac{F_T L}{AE} \frac{\partial F_T}{\partial S} = 0$$

But L, A and E are the same for all four members and from equation (i) we have

$$\frac{\partial F_T}{\partial R} = r \quad \text{and} \quad \frac{\partial F_T}{\partial S} = s$$

thus $\sum_1^4 (F + rR + sS)r = 0$

and $\sum_1^4 (F + rR + sS)s = 0$

or $\sum_1^4 Fr + R \sum_1^4 r^2 + S \sum_1^4 rs = 0 \qquad \text{(iii)}$

and $\sum_1^4 Fs + R \sum_1^4 rs + S \sum_1^4 s^2 = 0 \qquad \text{(iv)}$

Equations (iii) and (iv) are a pair of simultaneous equations for R and S. The coefficients of which are best obtained from a table showing the results of the force analysis of the systems shown in figure 2.5 - see table 2.1.

TABLE 2.1

| Member | F(kN) | r | s | $r^2$ | $s^2$ | rs | Fr | Fs |
|--------|-------|---|---|-------|-------|-----|-----|-----|
| OA | +10 | $-1/\sqrt{2}$ | $+1/\sqrt{2}$ | 1/2 | 1/2 | -1/2 | $-10/\sqrt{2}$ | $+10/\sqrt{2}$ |
| OB | 0 | 1 | 0 | 1 | 0 | 0 | 0 | 0 |
| OC | + 5 | $-1/\sqrt{2}$ | $-1/\sqrt{2}$ | 1/2 | 1/2 | +1/2 | $- 5/\sqrt{2}$ | $- 5/\sqrt{2}$ |
| OD | 0 | 0 | 1 | 0 | 1 | 0 | 0 | 0 |
|    |   |   | $\Sigma$ = | 2 | 2 | 0 | $-15/\sqrt{2}$ | $+ 5/\sqrt{2}$ |

Substitution of the summations from table 2.1 into equations (iii) and (iv) leads to

$$\frac{-15}{\sqrt{2}} + 2R = 0$$

and $\frac{+5}{\sqrt{2}} + 2S = 0$

thus $R = 15/2\sqrt{2}$ kN

and $S = -5/2\sqrt{2}$ kN

From the table and equation (i) it is possible to calculate the individual member forces as follows

$F_{OA} = +5$ kN

$F_{OB} = +15/2\sqrt{2}$ kN

$F_{OC} = +5/2$ kN

$F_{OD} = -5/2\sqrt{2}$ kN

the negative sign denotes compression.

Problems involving frames with non-linear elastic members may also be dealt with effectively as the next example shows.

*Example 2.2*

The three identical rods shown in figure 2.6 are pinned together at O and to rigid supports at A, B and C. The joint O carries a vertical load of 10 kN.

The rods are made from a non-linear elastic material which has a relationship between stress, $\sigma$, and strain, $\varepsilon$, given by

$$\sigma^3 = B\varepsilon$$

where B is a constant

Determine the forces in the rods.

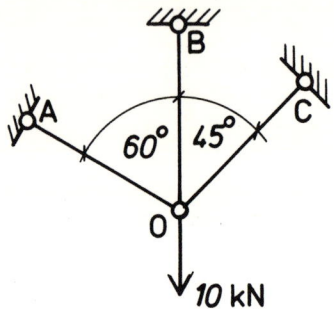

Figure 2.6

The frame has one redundancy and two degrees of freedom (the horizontal and vertical displacements of the joint O), thus the compatibility method of analysis is used here. (A solution by the equilibrium method is investigated in section 2.5.)

If OB is taken to be the redundant member, the frame may be converted into the two statically determinate systems shown in figure 2.7.

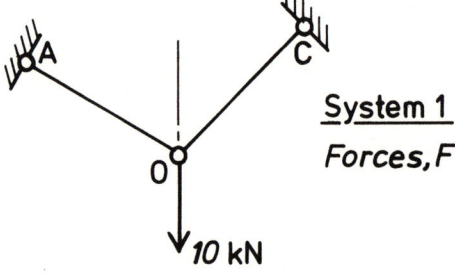

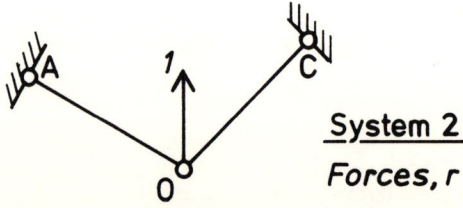

Figure 2.7

18

From equation 2.2 we have the complementary energy per unit volume as

$$c = \int_0^{\sigma_T} \varepsilon \, d\sigma$$

thus $c_{rod} = AL\int_0^{\sigma_T} \dfrac{\sigma^3}{B} d\sigma = \dfrac{\sigma_T^4}{4} \dfrac{AL}{B}$

since the volume of each rod is AL.

The complementary strain energy for the whole frame is therefore given by

$$C = \sum_1^3 \dfrac{F_T^4 L}{4A^3 B}$$

where $F_T = A\sigma_T$. Thus from the second theorem of complementary energy we have

$$\dfrac{\partial C}{\partial R} = \sum_1^3 \dfrac{F_T^3 L}{A^3 B} \dfrac{\partial F_T}{\partial R} = 0$$

where $F_T = F + rR$. Hence

$$\sum_1^3 (F + rR)^3 r = 0 \qquad (i)$$

Since L, A and B are the same for all members. Expanding equation (i), we obtain the following cubic equation for R

$$\sum_1^3 F^3 r + 3R\sum_1^3 F^2 r^2 + 3R^2 \sum_1^3 F r^3 + R^3 \sum_1^3 r^4 = 0 \qquad (ii)$$

The coefficients of this equation are obtained from table 2.2, which shows the results of the force analysis of the two statically determinate systems of figure 2.7.

TABLE 2.2

| Member | F | r | $F^3 r$ | $F^2 r^2$ | $F r^3$ | $r^4$ |
|---|---|---|---|---|---|---|
| OA | 7.320 | -0.732 | -287.2 | 28.72 | -2.872 | 0.2872 |
| OB | 0 | 1 | 0 | 0 | 0 | 1 |
| OC | 8.966 | -0.897 | -646.2 | 64.62 | -6.462 | 0.6462 |
|  |  | $\Sigma =$ | -933.4 | 93.34 | -9.334 | 1.9334 |

After substitution of the summations into equation (ii) and rearrangement, we have

$$R^3 - 14.483R^2 + 144.83R - 482.8 = 0 \qquad \text{(iii)}$$

Equation (iii) has only one real root and solution by trial gives

$$R = 4.94 \text{ kN}$$

thus $F_{OA} = +3.70$ kN

$F_{OB} = +4.94$ kN

$F_{OC} = +4.53$ kN

More general non-linear elastic problems usually lead to equations for the redundancies which are more difficult to solve than equation (iii) above. Had the above problem involved more than one redundancy for example, it would have been necessary to solve a set of simultaneous cubic equations.

It is of interest to compare the results of the force analysis for this problem with those which would have been obtained had the rod material been linearly elastic. The reader may wish to confirm that in this case the forces in the rods are

$F_{OA} = +3.13$ kN

$F_{OB} = +5.73$ kN

$F_{OC} = +3.83$ kN

The determination of the deflexions of the load point will be found in section 2.5 when the equilibrium approach to this problem is considered.

*Example 2.3*

Determine the forces in the frame shown in figure 2.2. The bar material is linearly elastic and the ratio of bar length to cross-sectional area is a constant.

The conversion of this frame into three statically determinate systems has already been discussed in section 2.2. The total member force is (see figure 2.3)

$$F_T = F + rR + sS \qquad \text{(i)}$$

where F is the force in a member due to external loading alone, R is the vertical reaction at C and S is the force in AD. r and s are the member forces produced by unit values of R and S respectively.

The complementary energy stored in the frame is obtained, for linearly elastic members, from equation 2.2

$$C = \sum_{1}^{7} \frac{F_T^2 L}{2AE} \qquad \text{(ii)}$$

By application of the second theorem of complementary energy (equation 1.5) we have

$$\frac{\partial \varepsilon}{\partial R} = \sum_{1}^{7} \frac{F_T L}{AE} \frac{\partial F_T}{\partial R} = 0 \qquad \text{(iii)}$$

and $\quad \dfrac{\partial \varepsilon}{\partial S} = \sum_{1}^{7} \dfrac{F_T L}{AE} \dfrac{\partial F_T}{\partial S} = 0 \qquad \text{(iv)}$

From equations (i), (iii) and (iv) and noting that L/AE is a constant we obtain

$$\sum_{1}^{7} Fr + R\sum_{1}^{7} r^2 + S\sum_{1}^{7} rs = 0 \qquad \text{(v)}$$

and $\quad \sum_{1}^{7} Fs + R\sum_{1}^{7} rs + S\sum_{1}^{7} s^2 = 0 \qquad \text{(vi)}$

The force analysis of the three statically determinate systems shown in figure 2.3 is straight forward. The results are given in table 2.3 together with the product terms.

TABLE 2.3

| Member | F(kN) | r | s | Fr | Fs | rs | $r^2$ | $s^2$ |
|---|---|---|---|---|---|---|---|---|
| AB | 5 | -2 | $-1/\sqrt{2}$ | -10 | $-5/\sqrt{2}$ | $\sqrt{2}$ | 4 | 1/2 |
| BC | 0 | $-\sqrt{2}$ | 0 | 0 | 0 | 0 | 2 | 0 |
| CD | 0 | 1 | 0 | 0 | 0 | 0 | 1 | 0 |
| DE | 0 | 1 | $-1/\sqrt{2}$ | 0 | 0 | $-1/\sqrt{2}$ | 1 | 1/2 |
| AD | 0 | 0 | 1 | 0 | 0 | 0 | 0 | 1 |
| BE | $-5\sqrt{2}$ | $\sqrt{2}$ | 1 | -10 | $-5\sqrt{2}$ | $\sqrt{2}$ | 2 | 1 |
| BD | 0 | 0 | $-1/\sqrt{2}$ | 0 | 0 | 0 | 0 | 1/2 |
| | | | $\Sigma =$ | -20 | $-15/\sqrt{2}$ | $3/\sqrt{2}$ | 10 | 7/2 |

Substitution of the summations into equations (v) and (vi) leads to the following simultaneous equations for R and S

$$10\sqrt{2}R + 3S = 20\sqrt{2} \qquad \text{(vii)}$$

and $\quad 3\sqrt{2}R + 7S = 15\sqrt{2} \qquad \text{(viii)}$

thus $R = \dfrac{95}{61}$ kN

and $\quad S = \dfrac{630\sqrt{2}}{427}$ kN

From equation (i) and the tabulated values of F, r and s we have

$F_{AB} = +0.41$ kN

$F_{BC} = -2.20$ kN

$F_{CD} = +1.56$ kN

$F_{DE} = +0.08$ kN

$F_{AD} = +2.09$ kN

$F_{BE} = -2.78$ kN

$F_{BD} = -1.47$ kN

*2.4.1 Lack of fit*

Problems involving a lack of fit produced by manufacturing errors or by deliberate mechanical means (the turnbuckle) may be solved by application of the second theorem of complementary energy (section 1.4).

*Example 2.4*

The cross-sectional area of each member of the truss in figure 2.8 is 500 mm$^2$ and the elastic modulus, E, is 50 GN m$^{-2}$. The turnbuckle along AC is tightened so that its two ends are brought 5 mm closer to each other. Determine the load in AC induced by this operation.

(Southampton)

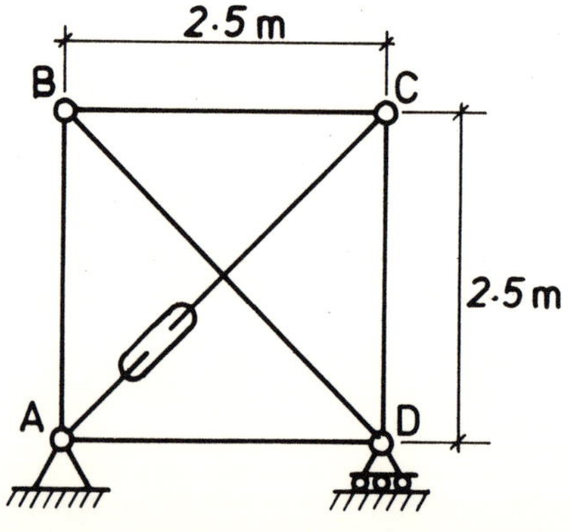

Figure 2.8

The effect of tightening the turnbuckle is to make AC too short to fit into the frame without inducing self-straining. The lack of fit, $\lambda$, of AC is therefore given by

$$\lambda = -0.005 \text{ m}$$

where the negative sign is in accordance with the convention adopted in section 1.4.

By reference to figure 1.4d it is possible to derive an expression for the complementary strain energy stored in a frame consisting of linearly elastic members subject to initial lack of fit, $\lambda$. Thus

$$C = \Sigma \left( \frac{F_T^2 L}{2AE} + F_T \lambda \right)$$

If AC in figure 2.8 is taken to be the redundant member, we have from the second theorem of complementary energy

$$\frac{\partial C}{\partial R} = \sum_1^6 \left( \frac{F_T L}{AE} + \lambda \right) \frac{\partial F_T}{\partial R} = 0 \tag{i}$$

where R is the force in AC.

There are no member forces due to external loads, thus

$$F_T = rR$$

therefore, since AE is a constant for all members, equation (i) becomes

$$R \sum_1^6 r^2 L + AE \sum_1^6 r\lambda = 0 \tag{ii}$$

where r is the force in a member produced by unit force in AC.

The values of r are calculated and shown in table 2.4 together with the products $r^2 L$ and $r\lambda$.

TABLE 2.4

| Member | r | L(m) | $\lambda$(m) | $r^2 L$(m) | $r\lambda$(m) |
|---|---|---|---|---|---|
| AB | $-1/\sqrt{2}$ | 2.5 | 0 | 1.25 | 0 |
| BC | $-1/\sqrt{2}$ | 2.5 | 0 | 1.25 | 0 |
| CD | $-1/\sqrt{2}$ | 2.5 | 0 | 1.25 | 0 |
| DA | $-1/\sqrt{2}$ | 2.5 | 0 | 1.25 | 0 |
| BD | 1 | $2.5\sqrt{2}$ | 0 | $2.5\sqrt{2}$ | 0 |
| AC | 1 | $2.5\sqrt{2}$ | -0.005 | $2.5\sqrt{2}$ | -0.005 |
|  |  |  | $\Sigma$ | $5(1+\sqrt{2})$ | -0.005 |

Substituting the summations into equation (ii) and noting that AE = 25000 kN, we have

$$5R(1 + \sqrt{2}) - 25{,}000 \times 0.005 = 0$$

thus $R = \dfrac{25}{(1 + \sqrt{2})} = 10.35$ kN

Had the turnbuckle been tightened still further so that C approached A by 5 mm, it would be possible to determine the new force in AC from equation (i) of section 1.4, for then

$$C_1 = \sum_1^5 \frac{F_T^2 L}{2AE}$$

where the summation excludes member AC. Thus

$$\frac{\partial C_1}{\partial R} = \sum_1^5 \frac{F_T L}{AE} \frac{\partial F_T}{\partial R} = \Delta_R$$

hence

$$R \sum_1^5 r^2 L = AE \Delta_R$$

The summation of the products $r^2 L$ may be obtained from the first five rows of table 2.4. Noting also that $\Delta_R = 0.005$ m, we have

$$2.5\sqrt{2}(1 + \sqrt{2})R = 25\,000 \times 0.005$$

hence

$$R = \frac{25\sqrt{2}}{(1 + \sqrt{2})} = 14.64 \text{ kN}$$

The reader may wish to verify that in the first case, where the lack of fit in AC was 5 mm, C would have approached A by 3.53 mm. Similarly an initial lack of fit of 7.07 mm in AC would account for C approaching A by 5 mm. The differences in the two sets of figures is due to the elastic extension of AC; this being 1.47 mm in the first case and 2.07 mm in the second.

Unfortunately, there is some ambiguity in this problem as stated. If the turnbuckle is closed by 5 mm before AC is inserted into the frame, then the first set of solutions applies. If, however, the member AC is initially joined to the frame with the turnbuckle loose and subsequently the latter is tightened until the distance between A and C is reduced by 5 mm, the second set of solutions applies.

The initial lack of fit need not necessarily be in the redundant member alone. Suppose that in example 2.4, all the members were too short by 5 mm, then

$$\sum_1^6 r\lambda = 0.010[\sqrt{(2)} - 1]\,\text{m}$$

and from equation (ii)
$$R = -\frac{50[\sqrt{2}-1]}{[\sqrt{2}+1]} = -8.58 \text{ kN}$$

*Example 2.5*

Determine the force in member AC of example 2.4 if all the members are made of a material whose stress-strain relationship is given by

$$\sigma^3 = B\varepsilon$$

where $B = 8 \times 10^{15}$ kN$^3$ m$^{-6}$.

The expression for the complementary strain energy may be obtained by reference to examples 2.2 and 2.4 as

$$C = \sum_1^6 \left( \frac{F_T^4 L}{4A^3 B} + F_T \lambda \right)$$

thus by application of the second theorem of complementary energy, we have

$$\frac{\partial C}{\partial R} = \sum_1^6 \left( \frac{F_T^3 L}{A^3 B} + \lambda \right) \frac{\partial F_T}{\partial R} = 0$$

hence

$$\frac{R^3}{A^3 B} \sum_1^6 r^4 L + \sum_1^6 r\lambda = 0 \tag{i}$$

since $F_T = rR$. Using the information in table 2.4 we have

$$\sum_1^6 r^4 L = \frac{5}{2}(1 + 2\sqrt{2}) \text{ m}$$

and $\sum_1^6 r\lambda = -0.005$ m as before

also $A^3 B = 10^6$ kN$^3$

Substituting these values in equation (i) we obtain

$$R^3 = \frac{2 \times 10^3}{(1 + 2\sqrt{2})} \text{ kN}^3$$

or $R = 8.05$ kN

The negative root may be rejected by inspection.

*Example 2.6*

The pin-jointed frame shown in figure 2.9 is composed of members

each having a cross-sectional area of 500 mm², the modulus of
elasticity being 200 GN m⁻². Find: (a) The force in member BD,
assuming that there is no initial lack of fit in any of the members,
for the particular loading shown.  (b) the initial lack of fit in
member BD that would double the force in BD found in (a).    (Leeds)

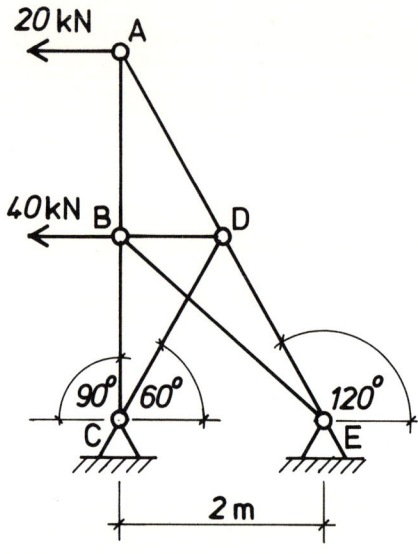

Figure 2.9

The frame has one redundant member and may therefore be converted
into the two statically determinate systems shown in figure 2.10.

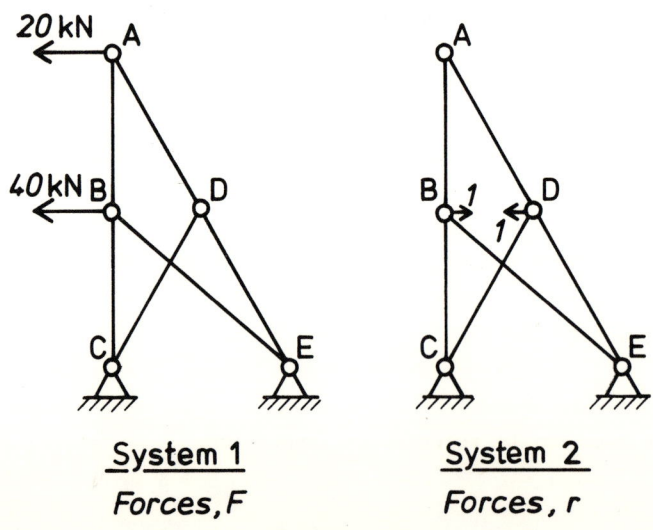

Figure 2.10

Assuming an initial lack of fit, $\lambda$, in member BD we have, from equation (i) of example 2.4

$$\sum_1^7 \left(\frac{F_T L}{AE} + \lambda\right)\frac{\partial F_T}{\partial R} = 0$$

where R is the force in BD, and

$$F_T = F + rR$$

thus $\sum_1^7 FrL + R\sum_1^7 r^2 L + AE\sum_1^7 r\lambda = 0$ \hfill (i)

Carrying out the force analysis for systems 1 and 2 in figure 2.10 and tabulating the relevant results, we have table 2.5.

TABLE 2.5

| Member | F(kN) | r | L(m) | $\lambda$(m) | FrL(kNm) | $r^2$L(m) | $r\lambda$(m) |
|---|---|---|---|---|---|---|---|
| AB | $-20\sqrt{3}$ | 0 | $\sqrt{3}$ | 0 | 0 | 0 | 0 |
| AD | 40 | 0 | 2 | 0 | 0 | 0 | 0 |
| BC | $-40\sqrt{3}$ | $\sqrt{(3)}/2$ | $\sqrt{3}$ | 0 | $-60\sqrt{3}$ | $3\sqrt{(3)}/4$ | 0 |
| BE | $20\sqrt{7}$ | $-\sqrt{(7)}/2$ | $\sqrt{7}$ | 0 | $-70\sqrt{7}$ | $7\sqrt{(7)}/4$ | 0 |
| DC | 0 | $-1$ | 2 | 0 | 0 | 2 | 0 |
| DE | 40 | 1 | 2 | 0 | 80 | 2 | 0 |
| BD | 0 | 1 | 1 | $\lambda$ | 0 | 1 | $\lambda$ |
| | | | | $\Sigma$ | $-209.12$ | $43.72$ | $\lambda$ |

Substitution of the summations into equation (i) and noting that $AE = 10^5$ kN gives

$$43.72R + 10^5 \lambda = 209.12 \text{ kN m}$$

For case (a) we have $\lambda = 0$, thus

$$R = 4.78 \text{ kN}$$

For case (b), $R = 9.56$ kN, thus

$$\lambda = -\frac{209.12}{10^5} = -0.0021 \text{ m}$$

$$= -2.1 \text{ mm}$$

The negative sign indicates that in case (b) member BD is initially too short.

*2.4.2 Temperature effects*

A rise or fall in temperature causing a change in the length of a member in a structural system will have no effect on the internal forces if the structure is statically determinate. A statically indeterminate structure, however, will usually suffer an alteration of its internal force system.

An exception to this rule concerns the statically indeterminate structure made from a single material which is statically determinate with respect to its supports, (the reactions being independent of the redundant forces). If such a structure is subjected to a uniform temperature change, no additional internal forces will be generated.

It is not difficult to see that this should be so, since all the members change their length in proportion. The temperature change simply applies a length scaling factor to the structure. Provided the supports are free to move in order to accommodate this overall change of size, no additional internal forces can be produced.

The temperature change in length of a bar in a pin-jointed frame may be conveniently treated as an initial lack of fit given by

$$\lambda = L\alpha\theta$$

where L is the original length of the bar, $\alpha$ is the temperature coefficient of expansion ($K^{-1}$) and $\theta$ is the temperature change (K).

*Example 2.7*

The pin-jointed plane frame ABCDEF shown in figure 2.11 consists of nine aluminium bars (E = 70 GN $m^{-2}$) each having a cross-sectional area of 120 $mm^2$ and a length of 2 m. A vertical load of 1 kN is applied at F.

If members AB, BC, CD and DE undergo an increase in temperature of 20 K relative to the other members, determine the force in member CF. The coefficient of linear expansion for aluminium is $23 \times 10^{-6}$ $K^{-1}$.

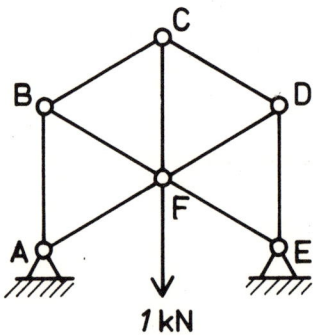

Figure 2.11

The frame has one redundancy which we will assume to be the horizontal thrust at E. Conversion of the statically indeterminate frame into two statically determinate systems is illustrated in figure 2.12.

**System 1**
Forces, F

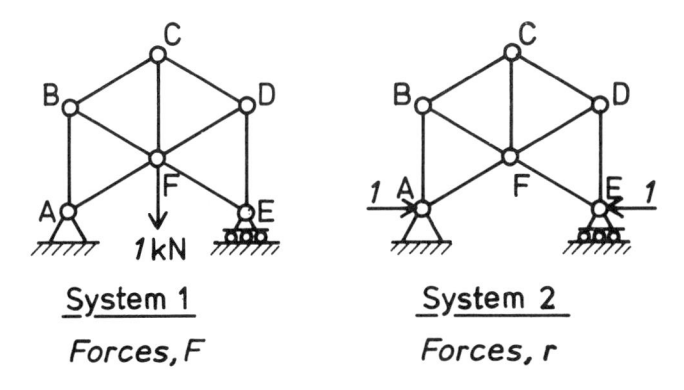

**System 2**
Forces, r

Figure 2.12

The complementary strain energy for a linearly elastic system with initial lacks of fit $\lambda$, is given by

$$C = \Sigma\left(\frac{F_T^2 L}{2AE} + F_T \lambda\right)$$

From the second complementary energy theorem (equation 1.5) we have

$$\frac{\partial C}{\partial R} = \Sigma\left(\frac{F_T L}{AE} + \lambda\right)\frac{\partial F_T}{\partial R} = 0 \qquad (i)$$

Now $F_T = F + rR$ and $L/AE$ = constant, thus from equation (i) we have

$$\sum_1^9 Fr + R\sum_1^9 r^2 + \frac{AE}{L}\sum_1^9 r\lambda = 0 \qquad (ii)$$

but $\lambda = L\alpha\theta$, thus equation (ii) becomes

$$\sum_1^9 Fr + R\sum_1^9 r^2 + \alpha AE\sum_1^9 r\theta = 0 \qquad (iii)$$

since all members have the same length and coefficient of expansion.

The force analysis of systems 1 and 2 in figure 2.12 gives forces F and r which are given in table 2.6 together with the product terms needed for equation (iii).

29

Table 2.6

| Member | F(kN) | r | θ (K) | Fr(kN) | $r^2$ | rθ (K) |
|---|---|---|---|---|---|---|
| AB | -1/2 | $1/\sqrt{3}$ | 20 | $-1/2\sqrt{3}$ | 1/3 | $20/\sqrt{3}$ |
| BC | -1/2 | $1/\sqrt{3}$ | 20 | $-1/2\sqrt{3}$ | 1/3 | $20/\sqrt{3}$ |
| CD | -1/2 | $1/\sqrt{3}$ | 20 | $-1/2\sqrt{3}$ | 1/3 | $20/\sqrt{3}$ |
| DE | -1/2 | $1/\sqrt{3}$ | 20 | $-1/2\sqrt{3}$ | 1/3 | $20/\sqrt{3}$ |
| AF | 0 | $-2/\sqrt{3}$ | 0 | 0 | 4/3 | 0 |
| BF | 1/2 | $-1/\sqrt{3}$ | 0 | $-1/2\sqrt{3}$ | 1/3 | 0 |
| CF | 1/2 | $-1/\sqrt{3}$ | 0 | $-1/2\sqrt{3}$ | 1/3 | 0 |
| DF | 1/2 | $-1/\sqrt{3}$ | 0 | $-1/2\sqrt{3}$ | 1/3 | 0 |
| EF | 0 | $-2/\sqrt{3}$ | 0 | 0 | 4/3 | 0 |
|  |  |  | Σ = | $-7/2\sqrt{3}$ | 5 | $80/\sqrt{3}$ |

Substituting the summations from table 2.6 into equation (iii) and noting that $\alpha AE = 0.1932$ kN K$^{-1}$, we have

$$-\frac{7}{2\sqrt{3}} + 5R + 0.1932 \times \frac{80}{\sqrt{3}} = 0$$

hence

$$R = -1.38 \text{ kN}$$

and $F_{CF} = \frac{1}{2} - \frac{1}{\sqrt{3}}(-1.38) = 1.30$ kN

*Example 2.8*

Determine the force in member DC of the plane, pin-jointed frame ABCDEF shown in figure 2.13. Find the new force in this member if the temperature of the whole frame is raised by 2 K. What temperature rise will result in zero force in member DC? All members are linearly elastic with AE/L = $10^5$ kN m$^{-1}$ and $\alpha = 12 \times 10^{-6}$ K$^{-1}$.

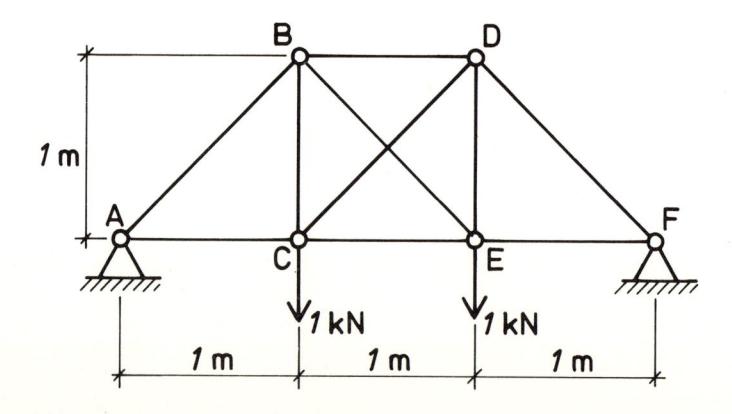

Figure 2.13

There are two redundant forces in this frame. We shall assume that they are the force R in DC and the horizontal reaction S at F. The three statically determinate systems to be considered are shown in figure 2.14.

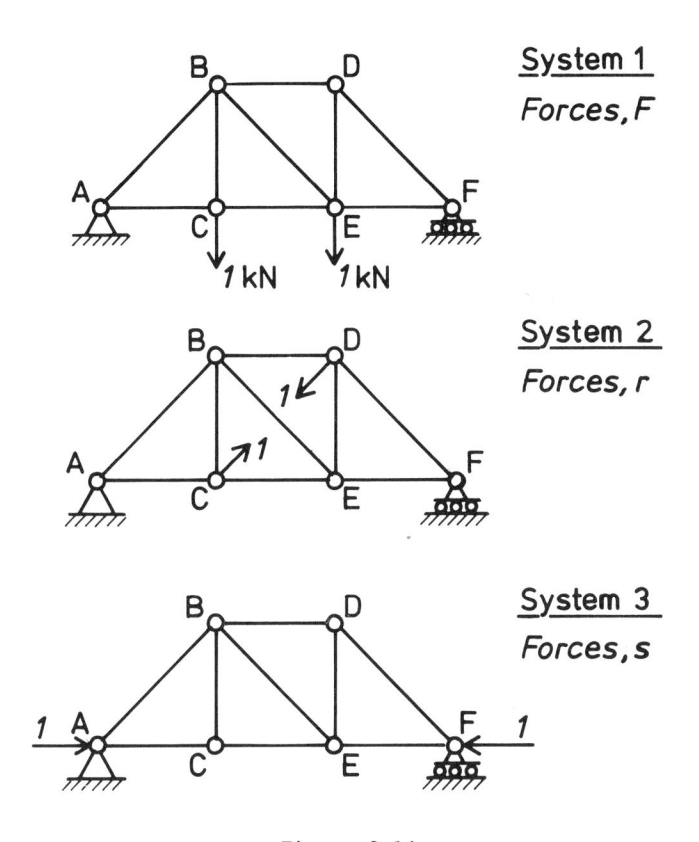

Figure 2.14

From the expression for the complementary strain energy given in example 2.8 and by application of the second complementary energy theorem, (equation 1.5) we obtain the governing equations for this problem as

$$\sum_{1}^{10} Fr + R\sum_{1}^{10} r^2 + S\sum_{1}^{10} rs + \frac{\alpha AE\theta}{L} \sum_{1}^{10} rL = 0 \tag{i}$$

and $\sum_{1}^{10} Fs + R\sum_{1}^{10} rs + S\sum_{1}^{10} s^2 + \frac{\alpha AE\theta}{L} \sum_{1}^{10} sL = 0$ (ii)

since $\lambda = L\alpha\theta$ and $\alpha$ and $\theta$ are the same for all members. Table 2.7 gives the results.

Table 2.7

| Member | F(kN) | r | s | L(m) | Fr(kN) | Fs(kN) | $r^2$ | $s^2$ | rs | rL(m) | sL(m) |
|---|---|---|---|---|---|---|---|---|---|---|---|
| AB | $-\sqrt{2}$ | 0 | 0 | $\sqrt{2}$ | 0 | 0 | 0 | 0 | 0 | 0 | 0 |
| AC | 1 | 0 | -1 | 1 | 0 | -1 | 0 | 1 | 0 | 0 | -1 |
| BD | -1 | $-1/\sqrt{2}$ | 0 | 1 | $1/\sqrt{2}$ | 0 | 1/2 | 0 | 0 | $-1/\sqrt{2}$ | 0 |
| BE | 0 | 1 | 0 | $\sqrt{2}$ | 0 | 0 | 1 | 0 | 0 | $\sqrt{2}$ | 0 |
| BC | 1 | $-1/\sqrt{2}$ | 0 | 1 | $-1/\sqrt{2}$ | 0 | 1/2 | 0 | 0 | $-1/\sqrt{2}$ | 0 |
| DC | 0 | 1 | 0 | $\sqrt{2}$ | 0 | 0 | 1 | 0 | 0 | $\sqrt{2}$ | 0 |
| CE | 1 | $-1/\sqrt{2}$ | -1 | 1 | $-1/\sqrt{2}$ | -1 | 1/2 | 1 | $1/\sqrt{2}$ | $-1/\sqrt{2}$ | -1 |
| DE | 1 | $-1/\sqrt{2}$ | 0 | 1 | $-1/\sqrt{2}$ | 0 | 1/2 | 0 | 0 | $-1/\sqrt{2}$ | 0 |
| DF | $-\sqrt{2}$ | 0 | 0 | $\sqrt{2}$ | 0 | 0 | 0 | 0 | 0 | 0 | 0 |
| EF | 1 | 0 | -1 | 1 | 0 | -1 | 0 | 1 | 0 | 0 | -1 |
|  |  |  |  | $\Sigma = $ | $-\sqrt{2}$ | -3 | 4 | 3 | $1/\sqrt{2}$ | 0 | -3 |

After substitution of the calculated quantities into equations (i) and (ii) and noting that $\alpha AE/L = 1.2$ kN m$^{-1}$ K$^{-1}$ we obtain

$$-\sqrt{2} + 4R + \frac{1}{\sqrt{2}}S = 0 \qquad (iii)$$

and 
$$-3 + \frac{1}{\sqrt{2}}R + 3S - 3.6\theta = 0 \qquad (iv)$$

Solving equations (iii) and (iv) for R alone, we have

$$R = \frac{3\sqrt{2}}{23} (1 - 1.2\theta) \text{ kN}$$

Now R is the force in member DC which we are required to find, thus

for $\theta = 0$, $R = \frac{3\sqrt{2}}{23} = 0.184$ kN

for $\theta = 2K$, $R = -\frac{4.2\sqrt{2}}{23} = -0.258$ kN

and for $R = 0$, $\theta = \frac{1}{1.2} = 0.833$ K

*2.4.3 Flexible Supports*

In practice, no structural support is truely rigid. It is often necessary to take account of this fact by considering the supports as stiff springs. When carrying out the force analysis of a system with flexible supports these springs may be treated as additional members. Two examples will suffice to illustrate the approach.

*Example 2.9*

The pin-jointed structure shown in figure 2.15 carries a vertical load load of 80 kN at C. The structure is hinged to unyielding supports at A and D. Additional support is given to the structure by the roller bearing at E, but this support yields 0.01 mm/kN reaction.

Determine the forces in members AE and DE for the given loading.
For all members, $AE/L = 6 \times 10^4$ kN m$^{-1}$.  (Strathclyde)

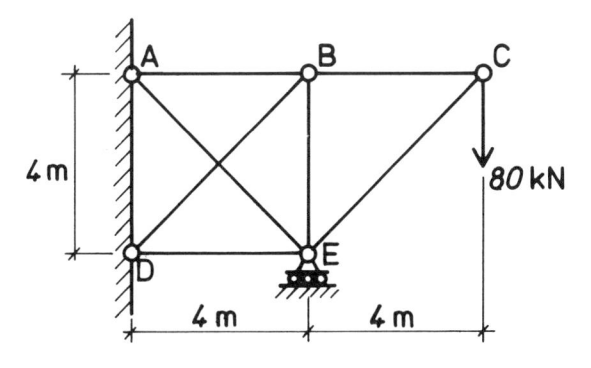

Figure 2.15

The force R in member AE and the vertical reaction S provided by the flexible support at E will be taken as the redundant forces for the frame. The three statically determinate systems to be considered are shown in figure 2.16.

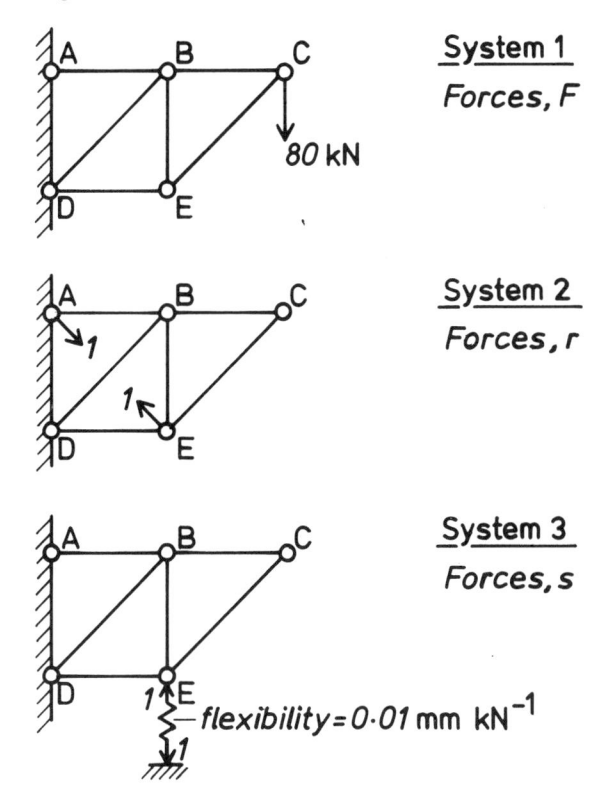

Figure 2.16

33

The support at E is treated as an additional member of flexibility $1/100$ mm $kN^{-1}$. The flexibility, $L/AE$, of all other members is given as $1/60$ mm $kN^{-1}$.

The total force, $F_T$, in all the members including the support spring is given by

$$F_T = F + rR + sS$$

All members are linearly elastic so that the complementary energy for the system is given by

$$C = \sum_1^8 \left(\frac{F_T^2 L}{2AE}\right)$$

where the summation includes the flexible support at E. Since $\partial C/\partial R = \partial C/\partial S = 0$, $\partial F_T/\partial R = r$, and $\partial F_T/\partial S = s$, we have, as the governing equations for the problem

$$\sum_1^8 \frac{FrL}{AE} + R\sum_1^8 \frac{r^2 L}{AE} + S\sum_1^8 \frac{rsL}{AE} = 0 \tag{i}$$

and

$$\sum_1^8 \frac{FsL}{AE} + R\sum_1^8 \frac{rsL}{AE} + S\sum_1^8 \frac{s^2 L}{AE} = 0 \tag{ii}$$

Table 2.8 shows the results of the force analysis of the three systems in figure 2.16 together with the product terms needed for equations (i) and (ii).

Table 2.8

| Member | F(kN) | r | s | $L/AE \left(\frac{mm}{kN}\right)$ | $\frac{FrL}{AE}$ (mm) | $\frac{FsL}{AE}$ (mm) | $\frac{r^2 L}{AE}\left(\frac{mm}{kN}\right)$ | $\frac{s^2 L}{AE}\left(\frac{mm}{kN}\right)$ | $\frac{rsL}{AE}\left(\frac{mm}{kN}\right)$ |
|---|---|---|---|---|---|---|---|---|---|
| AB | +160 | $-1/\sqrt{2}$ | $-1$ | 1/60 | $-4\sqrt{2}/3$ | $-8/3$ | 1/120 | 1/60 | $1/60\sqrt{2}$ |
| BD | $-80\sqrt{2}$ | 1 | $\sqrt{2}$ | 1/60 | $-4\sqrt{2}/3$ | $-8/3$ | 1/60 | 1/30 | $\sqrt{2}/60$ |
| DE | $-80$ | $-1/\sqrt{2}$ | 0 | 1/60 | $+2\sqrt{2}/3$ | 0 | 1/120 | 0 | 0 |
| AE | 0 | 1 | 0 | 1/60 | 0 | 0 | 1/60 | 0 | 0 |
| BE | +80 | $-1/\sqrt{2}$ | $-1$ | 1/60 | $-2\sqrt{2}/3$ | $-4/3$ | 1/120 | 1/60 | $1/60\sqrt{2}$ |
| CE | $-80\sqrt{2}$ | 0 | 0 | 1/60 | 0 | 0 | 0 | 0 | 0 |
| BC | +80 | 0 | 0 | 1/60 | 0 | 0 | 0 | 0 | 0 |
| Support, E | 0 | 0 | 1 | 1/100 | 0 | 0 | 0 | 1/100 | 0 |
| | | | | $\Sigma =$ | $-8\sqrt{2}/3$ | $-20/3$ | $+7/120$ | $+23/300$ | $+\sqrt{2}/30$ |

Substituting the summations into equations (i) and (ii) we have

$$-\frac{8\sqrt{2}}{3} + R\frac{7}{120} + S\frac{\sqrt{2}}{30} = 0$$

and

$$-\frac{20}{3} + \frac{R\sqrt{2}}{30} + S\frac{23}{300} = 0$$

From which

$$R = -\frac{4480\sqrt{2}}{567} = -11.17 \text{ kN}$$

and $\quad S = \frac{7600}{81} = +93.83 \text{ kN}$

Hence

$$F_{ae} = R = -11.17 \text{ kN}$$

and $\quad F_{de} = -80 - \frac{R}{\sqrt{2}} = -72.1 \text{ kN}$

As an extension of this problem, the reader is invited to show that: (a) if E is a rigid support, R = -37.7 kN and S = +126.7 kN. (b) For no support at E, R = +64.65 kN and S = 0. (c) If the flexibility of the support at E is relaxed to 1/60 mm kN$^{-1}$, R = 0 and S = 80 kN.

*Example 2.10*

The extremity of the cantilever bracket shown in figure 2.17a carries a vertical load of 20 kN. In order to limit the deflexion of the load point, support is provided in the form of a vertical rod joining the load point to the tip of a cantilever beam of length 1.2 m and flexural rigidity EI. The beam, the members of the bracket and the vertical rod are all of steel and have the same cross-sectional area. Determine the force in the vertical rod if the radius of gyration of the beam cross-section is 0.2 m.

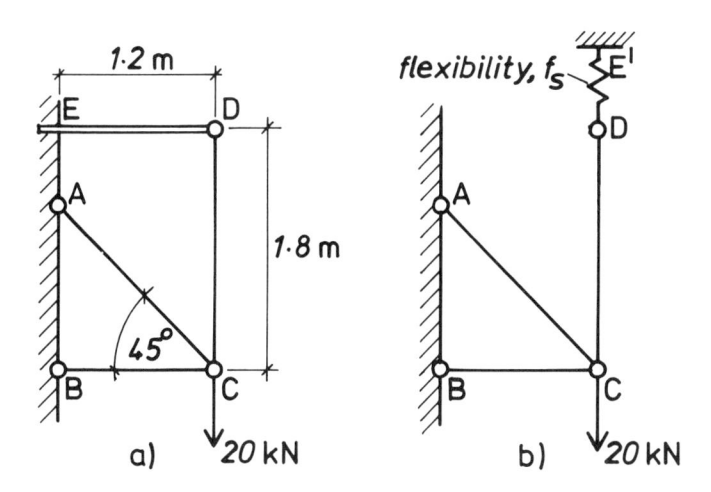

Figure 2.17

The cantilever beam provides flexible support to the top of the vertical rod and can be treated as the linear spring DE' shown in figure 2.17b.

The relationship between end load, R and end deflexion, $\Delta$ for a cantilever beam is given by (see Essential Solid Mechanics)

$$\Delta = \frac{RL^3}{3EI}$$

where L is the span and EI is the flexural rigidity. Thus the flexibility $f_s$ of the equivalent spring DE' in figure 2.17b is

$$f_s = \frac{L^3}{3EI}$$

But L = 1.2 m and I = (cross-sectional area) × (radius of gyration)$^2$, thus

$$f_s = \frac{(1.2)^3}{3AE(0.2)^2} = \frac{14.4}{AE} \text{ m kN}^{-1}$$

if AE is in kN.

The equivalent system in figure 2.17b has one redundant force which we will take to be the force, R in the spring. The two statically determinate systems to be considered are shown in figure 2.18.

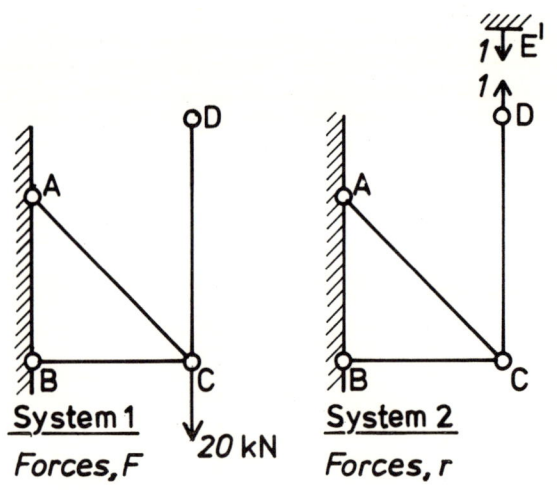

**System 1**
Forces, F

**System 2**
Forces, r

20 kN

Figure 2.18

By now the reader should be familiar with the setting up of the governing equation for this type of problem, which is found to be.

$$\sum_1^4 Frf + R\sum_1^4 r^2 f = 0 \qquad (i)$$

where f is the member flexibility.

Table 2.9 gives the results of the force analysis for the two systems in figure 2.18, together with the product terms.

Table 2.9

| Member | F (kN) | r | L (m) | f (m/kN) | Frf (m) | $r^2f$ (m/kN) |
|---|---|---|---|---|---|---|
| AC | $20\sqrt{2}$ | $-\sqrt{2}$ | $1.2\sqrt{2}$ | $1.2\sqrt{2}/AE$ | $-48\sqrt{2}/AE$ | $2.4\sqrt{2}/AE$ |
| BC | $-20$ | 1 | 1.2 | $1.2/AE$ | $-24/AE$ | $1.2/AE$ |
| DC | 0 | 1 | 1.8 | $1.8/AE$ | 0 | $1.8/AE$ |
| DE' | 0 | 1 | – | $14.4/AE$ | 0 | $14.4/AE$ |
| | | | | $\Sigma \;=\;$ | $-\dfrac{24}{AE}(1+2\sqrt{2})$ | $\dfrac{(17.4+2.4\sqrt{2})}{AE}$ |

After substituting the summations into equation (i) we obtain

$$R = \frac{24(1 + 2\sqrt{2})}{17.4 + 2.4\sqrt{2}} = 4.42 \text{ kN}$$

2.5 THE EQUILIBRIUM METHOD

At the beginning of section 2.4 it was pointed out that most redundant plane frames are more efficiently dealt with by the compatibility approach since there are usually more degrees of freedom in the system than redundant forces. There are, however, a few examples of redundant plane frames where this is not so and for these frames the equilibrium method may usefully be applied.

The first step in the analysis of a frame by the compatibility method was the determination of an equilibrium set of forces. In an analogous fashion, when applying the equilibrium approach, we start by finding a compatible set of member and joint displacements.

Example 2.1 will now be solved by the equilibrium approach to allow the reader to compare the two methods.

*Example 2.11*

Solve example 2.1 by the equilibrium method.

The system (see figure 2.4) has two degrees of freedom, the vertical and horizontal displacements of the load point O. Figure 2.19a shows the system subjected to the horizontal displacement ($\Delta_H$) alone. In figure 2.19b the system is subjected to the vertical displacement ($\Delta_V$) alone.

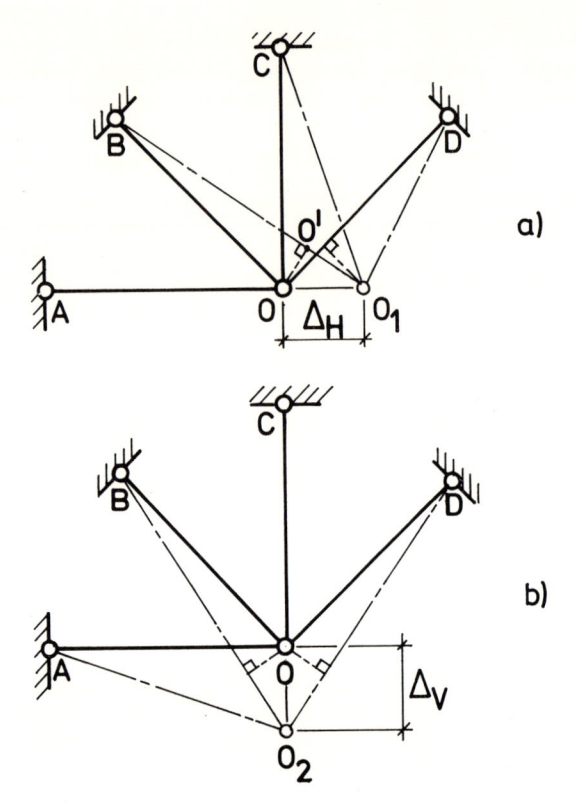

Figure 2.19

For small deflexions, the extension of member BO due to the single joint displacement, $\Delta_H$ is $O'O_1$. Since triangle $OO'O_1$ is approximately right-angled at $O'$ we have

$$\Delta_{BO} = O'O_1 = \Delta_H \cos 45°$$

Similarly

$$\Delta_{DO} = -\Delta_H \cos 45°$$

where the negative sign denotes a contraction.

The extension of AO is clearly equal to $\Delta_H$ while CO suffers no change in length provided $\Delta_H$ is small.

These results are recorded in table 2.10 together with those obtained by applying the single joint displacement $\Delta_V$.

Table 2.10

|  | $\Delta_{AO}$ | $\Delta_{BO}$ | $\Delta_{CO}$ | $\Delta_{DO}$ |
|---|---|---|---|---|
| $\Delta_H$ alone | $\Delta_H$ | $\Delta_H/\sqrt{2}$ | 0 | $-\Delta_H/\sqrt{2}$ |
| $\Delta_V$ alone | 0 | $\Delta_V/\sqrt{2}$ | $\Delta_V$ | $\Delta_V/\sqrt{2}$ |
| Total ($\Delta_T$) | $\Delta_H$ | $\dfrac{\Delta_H + \Delta_V}{\sqrt{2}}$ | $\Delta_V$ | $\dfrac{\Delta_V - \Delta_H}{\sqrt{2}}$ |

The equilibrium method makes use of Castigliano's first theorem, part I, (equation 1.1) in which the strain energy must be expressed in terms of displacements.

An expression for the strain energy in each rod may be determined from the stress-strain relationship. The strain-energy per unit volume (u) is obtained from equation 2.1 as

$$u = \int_0^{\varepsilon_r} \sigma \, d\varepsilon$$

For a linear elastic material ($\sigma = E\varepsilon$) this becomes

$$u = E \int_0^{\varepsilon_r} \varepsilon \, d\varepsilon$$

or $\quad u = E \dfrac{\varepsilon^2_T}{2} = \dfrac{E\Delta^2_T}{2L^2}$

Since the volume of each rod is AL, the strain energy for the whole frame is given by

$$U = \sum_1^4 \dfrac{EA\Delta_T^2}{2L} \qquad (i)$$

Applying Castigliano's first theorem, part I, we have

$$\dfrac{\partial U}{\partial \Delta_H} = \sum_1^4 \dfrac{EA\Delta_T}{L} \dfrac{\partial \Delta_T}{\partial \Delta_H} = 10 \text{ kN}$$

and $\quad \dfrac{\partial U}{\partial \Delta_V} = \sum_1^4 \dfrac{EA\Delta_T}{L} \dfrac{\partial \Delta_T}{\partial \Delta_V} = 5 \text{ kN}$

But L, A and E are the same for all four members thus

$$\sum_1^4 \Delta_T \dfrac{\partial \Delta_T}{\partial \Delta_H} = \dfrac{10L}{EA} \qquad (ii)$$

and $\sum_{1}^{4} \Delta_T \dfrac{\partial \Delta T}{\partial \Delta_V} = \dfrac{5L}{EA}$ (iii)

It should be noted that the right-hand sides of equations (ii) and (iii) are in units of length. To evaluate the summations, the form of calculation shown in table 2.11 is adopted.

Table 2.11

| Member | $\Delta_T$ | $\dfrac{\partial \Delta_T}{\partial \Delta_H}$ | $\dfrac{\partial \Delta_T}{\partial \Delta_V}$ | $\Delta_T \dfrac{\partial \Delta_T}{\partial \Delta_V}$ | $\Delta_T \dfrac{\partial \Delta_T}{\partial \Delta_V}$ |
|---|---|---|---|---|---|
| AO | $\Delta_H$ | 1 | 0 | $\Delta_H$ | 0 |
| BO | $\dfrac{\Delta_H + \Delta_V}{\sqrt{2}}$ | $1/\sqrt{2}$ | $1/\sqrt{2}$ | $\dfrac{\Delta_H + \Delta_V}{2}$ | $\dfrac{\Delta_H + \Delta_V}{2}$ |
| CO | $\Delta_V$ | 0 | 1 | 0 | $\Delta_V$ |
| DO | $\dfrac{\Delta_V - \Delta_H}{\sqrt{2}}$ | $-1/\sqrt{2}$ | $1/\sqrt{2}$ | $\dfrac{-\Delta_V + \Delta_H}{2}$ | $\dfrac{\Delta_V - \Delta_H}{2}$ |
|  |  |  | $\Sigma =$ | $2\Delta_H$ | $2\Delta_V$ |

After substituting the summations into equations (ii) and (iii) we obtain the deformations

$\Delta_H = \dfrac{5L}{AE}$ and $\Delta_V = \dfrac{5L}{2AE}$

If we require the numerical values of these deformations, L, A and E must be known, but if it is only the member forces that are required, this information is not necessary. To calculate the individual forces in the members, the member strains are first calculated from

$\varepsilon = \dfrac{\Delta_T}{L}$

then, since $\sigma = E\varepsilon$ and $F_T = \sigma A$, we have

$F_T = \dfrac{AE\Delta_T}{L}$

From the results above

$F_{OA} = +5$ kN

$F_{OB} = \dfrac{1}{\sqrt{2}} (5 + \dfrac{5}{2}) = +15/2\sqrt{2}$ kN

$F_{OC} = +5/2$ kN

$$F_{OD} = \frac{1}{\sqrt{2}} \left(\frac{5}{2} - 5\right) = -5/2\sqrt{2} \text{ kN}$$

the negative sign denotes compression.

To show how the equilibrium method can be applied to non-linear elastic problems, example 2.2 will now be re-worked.

*Example 2.12*

Solve example 2.2 by the equilibrium method.

Referring to figure 2.6, we see that if the common joint O is moved an amount $\Delta_H$ to the right and $\Delta_V$ downwards, the total member deformations ($\Delta_T$) are given by

$$(\Delta_T)_{OA} = (\Delta_V + \sqrt{(3)}\Delta_H)/2$$

$$(\Delta_T)_{OB} = \Delta_V$$

$$(\Delta_T)_{OC} = (\Delta_V - \Delta_H)/\sqrt{2}$$

The strain energy per unit volume is obtained from

$$u = \int_0^{\varepsilon_T} \sigma \, d\varepsilon$$

since $\sigma^3 = B\varepsilon$, we have

$$u = \int_0^{\varepsilon_T} B^{1/3} \varepsilon^{1/3} \, d\varepsilon = \frac{3B^{1/3}}{4} \varepsilon_T^{4/3}$$

or $$u = \frac{3B^{1/3}}{4} \left(\frac{\Delta_T}{L}\right)^{4/3}$$

Hence the total strain energy in the frame is given by

$$U = \sum_1^3 \frac{3AB^{1/3}}{4} \frac{\Delta_T^{4/3}}{L^{1/3}} \tag{i}$$

Applying Castigliano's first theorem, part I, we have

$$\frac{\partial U}{\partial \Delta_H} = \sum_1^3 A \left(\frac{B\Delta_T}{L}\right)^{1/3} \frac{\partial \Delta_T}{\partial \Delta_H} = 0$$

and $$\frac{\partial U}{\partial \Delta_V} = \sum_1^3 A \left(\frac{B\Delta_T}{L}\right)^{1/3} \frac{\partial \Delta_T}{\partial \Delta_V} = 10 \text{ kN}$$

Since A, B and L are the same for all three members, we have

$$\sum_{1}^{3} \Delta_T^{1/3} \frac{\partial \Delta_T}{\partial \Delta_H} = 0 \qquad \text{(ii)}$$

and

$$\sum_{1}^{3} \Delta_T^{1/3} \frac{\partial \Delta_T}{\partial \Delta_V} = \frac{10}{A} \left(\frac{L}{B}\right)^{1/3} \qquad \text{(iii)}$$

The partial differentials of $\Delta_T$ are shown in table 2.12.

Table 2.12

| Member | $\Delta_T$ | $\frac{\partial \Delta_T}{\partial \Delta_H}$ | $\frac{\partial \Delta_T}{\partial \Delta_V}$ |
|---|---|---|---|
| AO | $(\Delta_V + \sqrt{3}\Delta_H)/2$ | $\sqrt{3}/2$ | $1/2$ |
| BO | $\Delta_V$ | $0$ | $1$ |
| CO | $(\Delta_V - \Delta_H)/\sqrt{2}$ | $-1/\sqrt{2}$ | $1/\sqrt{2}$ |

Substitution in equation (ii) gives

$$\left(\frac{\Delta_V + \sqrt{3}\Delta_H}{2}\right)^{1/3} \frac{\sqrt{3}}{2} - \left(\frac{\Delta_V - \Delta_H}{\sqrt{2}}\right)^{1/3} \frac{1}{\sqrt{2}} = 0$$

hence

$$(\Delta_V + \sqrt{3}\Delta_H) \frac{3\sqrt{3}}{4} = (\Delta_V - \Delta_H)$$

or

$$\Delta_H = \frac{(4 - 3\sqrt{3})}{13} \Delta_V$$

The total member deflexions may now be obtained in terms of $\Delta_V$ as follows

$$(\Delta_T)_{OA} = 2(1 + \sqrt{3})\Delta_V/13$$

$$(\Delta_T)_{OB} = \Delta_V$$

$$(\Delta_T)_{OC} = 3\sqrt{3}(1 + \sqrt{3})\Delta_V/13\sqrt{2}$$

After substitution in equation (iii) we obtain

$$\Delta_V^{1/3} \left\{ \frac{1}{2} \left[\frac{2(1 + \sqrt{3})}{13}\right]^{1/3} + 1 + \frac{1}{\sqrt{2}} \left[\frac{3\sqrt{3}(1 + \sqrt{3})}{13\sqrt{2}}\right]^{1/3} \right\} = \frac{10}{A} \left(\frac{L}{B}\right)^{1/3}$$

or $\Delta_V \left[ 1 + (2 + \sqrt{3}) \left( \frac{1 + \sqrt{3}}{52} \right)^{1/3} \right]^3 = \frac{1000}{A} \left( \frac{L}{B} \right)$

hence

$$\Delta_V = \frac{120.73L}{A^3 B}$$

and $\Delta_H = -\frac{11.11L}{A^3 B}$

To obtain the member forces, the individual strains are first obtained from

$$\varepsilon = \frac{\Delta_T}{L}$$

then since the stress-strain relationship is given by

$\sigma^3 = B\varepsilon$ and $\sigma = F_T/A$ we have

$$F_T = A \left( \frac{B \Delta_T}{L} \right)^{1/3}$$

thus $F_{OA} = \left( \frac{120.73 - 11.11\sqrt{3}}{2} \right)^{1/3} = 3.70$ kN

$F_{OB} = (120.73)^{1/3} = 4.94$ kN

$F_{OC} = \left( \frac{120.73 + 11.11}{\sqrt{2}} \right)^{1/3} = 4.53$ kN

The deformations of the load point O are downwards and to the left. It is of interest to note that if the rod had been made of linear elastic material, these deformations would have been downwards and to the right. The reader is invited to confirm that, in this case

$$\Delta_H = \frac{0.31L}{AE}$$

and $\Delta_V = \frac{5.73L}{AE}$

## 2.5.1 Lack of fit and temperature effects

The equilibrium method may also be used to deal with problems involving a lack of fit or temperature effects.

Figure 2.20 shows the force-deformation relationship for a linear-elastic, axially loaded bar of nominal length L and lack of fit $\lambda$.

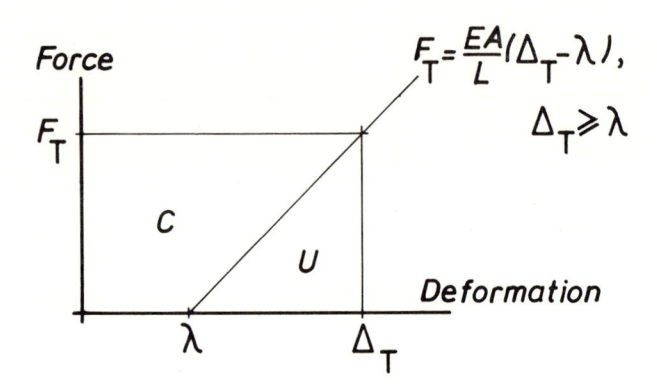

Figure 2.20

The strain energy (U) stored in the bar is represented by the area under the force-deformation curve. From Figure 2.20 we have

$$U = \tfrac{1}{2} F_T(\Delta_T - \lambda)$$

or $\quad U = \dfrac{EA}{2L}(\Delta_T - \lambda)^2 \qquad (2.3)$

This expression could have been derived from the integral for strain energy per unit volume (see example 2.11), the upper limit of integration being $\varepsilon_T = (\Delta_T - \lambda)/L$.

The following example illustrates the method of solution for this type of problem.

*Example 2.13*

A plane truss consisting of three pin-ended bars all of equal cross-sectional area, modulus of elasticity 200 GN m$^{-2}$ and coefficient of thermal expansion $10 \times 10^{-6}$ K$^{-1}$ as shown in figure 2.21, is subjected to a uniform temperature rise of 100 K.

Assuming that supports for the pin-joints at B, C and D are infinitely rigid, calculate the stresses induced in the bars and the displacements of point A. (Brunel)

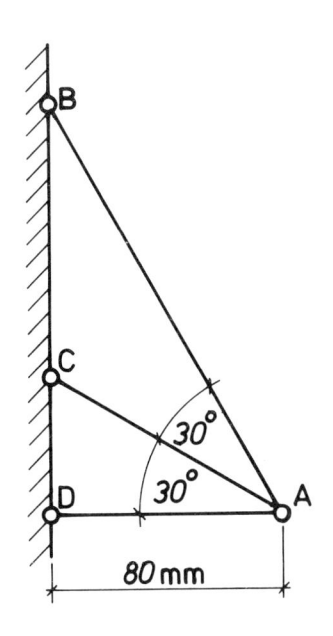

Figure 2.21

Let the horizontal and vertical displacements of point A be $\Delta_H$ and $\Delta_V$ respectively. The individual bar deformations are obtained in terms of these displacements and entered in table 2.13 together with the total deformations, $\Delta_T$.

Table 2.13

|  | $\Delta_{ba}$ | $\Delta_{ca}$ | $\Delta_{da}$ |
|---|---|---|---|
| $\Delta_H$ alone | $\Delta_H/2$ | $\sqrt{(3)}\Delta_H/2$ | $\Delta_H$ |
| $\Delta_V$ alone | $\sqrt{(3)}\Delta_V/2$ | $\Delta_V/2$ | 0 |
| $\Delta_T$ | $\dfrac{\Delta_H + \sqrt{(3)}\Delta_V}{2}$ | $\dfrac{\sqrt{(3)}\Delta_H + \Delta_V}{2}$ | $\Delta_H$ |

From equation 2.3 above and Castigliano's first theorem, Part I, (equation 1.1) we have

$$\frac{\partial U}{\partial \Delta_H} = \sum_1^3 \frac{EA}{L}(\Delta_T - \lambda)\frac{\partial \Delta_T}{\partial \Delta_H} = 0 \qquad (i)$$

and $\quad \dfrac{\partial U}{\partial \Delta_V} = \sum_1^3 \dfrac{EA}{L}(\Delta_T - \lambda)\dfrac{\partial \Delta_T}{\partial \Delta_V} = 0 \qquad (ii)$

Since $\lambda$ is independent of $\Delta_H$ and $\Delta_V$ and there are no external forces acting at A.

Table 2.14

| Member | $\Delta_T$ | $L$ (mm) | $\dfrac{\partial \Delta_T}{\partial \Delta_V}$ | $\dfrac{\partial \Delta_T}{\partial \Delta_V}$ | $\dfrac{\Delta_T}{L}$ | $\dfrac{\Delta_T}{L}\dfrac{\partial \Delta_T}{\partial \Delta_H}$ | $\dfrac{\Delta_T}{L}\dfrac{\partial \Delta_T}{\partial \Delta_V}$ |
|---|---|---|---|---|---|---|---|
| BA | $\dfrac{\Delta_H + \sqrt{3}\Delta_V}{2}$ | 160 | $\dfrac{1}{2}$ | $\dfrac{\sqrt{3}}{2}$ | $\dfrac{\Delta_H + \sqrt{3}\Delta_V}{320}$ | $\dfrac{\Delta_H + \sqrt{3}\Delta_V}{640}$ | $\dfrac{\sqrt{3}\Delta_H + 3\Delta_V}{640}$ |
| CA | $\dfrac{\sqrt{3}\Delta_H + \Delta_V}{2}$ | $\dfrac{160}{\sqrt{3}}$ | $\dfrac{\sqrt{3}}{2}$ | $\dfrac{1}{2}$ | $\dfrac{3\Delta_H + \sqrt{3}\Delta_V}{320}$ | $\dfrac{3\sqrt{3}\Delta_H + 3\Delta_V}{640}$ | $\dfrac{3\Delta_H + \sqrt{3}\Delta_V}{640}$ |
| DA | $\Delta_H$ | 80 | 1 | 0 | $\dfrac{\Delta_H}{80}$ | $\dfrac{\Delta_H}{80}$ | 0 |
| | | | $\Sigma = \dfrac{\sqrt{3}}{2}(1+\sqrt{3})$ | $\Sigma = \dfrac{1}{2}(1+\sqrt{3})$ | | $\Sigma = \dfrac{\sqrt{3}(1+\sqrt{3})}{640} \times (3\Delta_H + \Delta_V)$ | $\dfrac{\sqrt{3}(1+\sqrt{3})}{640} \times (\Delta_H + \Delta_V)$ |

The change in length of each member caused by the increase in temperature may be interpreted as a lack of fit $\lambda$, where $\lambda = L\alpha\theta$. The coefficient of thermal expansion ($\alpha$) and the temperature change ($\theta$) are given, thus for each member

$$\lambda = L \times 10^{-3}$$

Substituting for $\lambda$ in equations (i) and (ii) and noting that EA is the same for all members, we have

$$\sum_1^3 \frac{\Delta_T}{L} \frac{\partial \Delta_T}{\partial \Delta_H} - 10^{-3} \sum_1^3 \frac{\partial \Delta_T}{\partial \Delta_H} = 0 \qquad \text{(iii)}$$

and $\sum_1^3 \frac{\Delta_T}{L} \frac{\partial \Delta_T}{\partial \Delta_V} - 10^{-3} \sum_1^3 \frac{\partial \Delta_T}{\partial \Delta_V} = 0 \qquad \text{(iv)}$

The summations in these equations may be obtained from table 2.14.

Substitution from the table into equations (iii) and (iv) gives

$$\frac{\sqrt{3}(1 + \sqrt{3})}{640} (3\Delta_H + \Delta_V) = 10^{-3} \times \frac{\sqrt{3}}{2} (1 + \sqrt{3})$$

and $\frac{\sqrt{3}(1 + \sqrt{3})}{640} (\Delta_H + \Delta_V) = 10^{-3} \times \frac{1}{2} (1 + \sqrt{3})$

which simplify to

$$3\Delta_H + \Delta_V = 0.32 \text{ mm} \qquad \text{(v)}$$

and $\Delta_H + \Delta_V = \frac{0.32}{\sqrt{3}} \text{ mm} \qquad \text{(vi)}$

hence

$$\Delta_H = 0.068 \text{ mm}$$

and $\Delta_V = 0.117 \text{ mm}$

The member forces are obtained from

$$F_T = \frac{EA}{L}(\Delta_T - \lambda)$$

thus the member stresses are given by

$$\sigma_T = \frac{F_T}{A} = E\left(\frac{\Delta_T}{L} - \alpha\theta\right)$$

or $\sigma_T = 200 \times 10^3 \left(\frac{\Delta_T}{L} - 10^{-3}\right)$ MN m$^{-2}$

Evaluating the strains, $\Delta_T/L$ from table 2.14 we have

For BA: $\dfrac{\Delta_T}{L} = 2\left(1 - \dfrac{1}{\sqrt{3}}\right) \times 10^{-3}$

For CA: $\dfrac{\Delta_T}{L} = \sqrt{(3)}(\sqrt{3} - 1) \times 10^{-3}$

For DA: $\dfrac{\Delta_T}{L} = 2\left(1 - \dfrac{1}{\sqrt{3}}\right) \times 10^{-3}$

Hence

$$\sigma_{T\ BA} = 200\left(1 - \dfrac{2}{\sqrt{3}}\right) = -30.9 \text{ MN m}^{-2}$$

$$\sigma_{T\ CA} = 200(2 - \sqrt{3}) = +53.6 \text{ MN m}^{-2}$$

$$\sigma_{T\ DA} = 200\left(1 - \dfrac{2}{\sqrt{3}}\right) = -30.9 \text{ MN m}^{-2}$$

Non-linear elastic frames incorporating lack of fit of a member or members present no difficulties, as the next example shows.

*Example 2.14*

Determine the amount by which member CO in the frame of example 2.2 (figure 2.6) must be shortened so that, on loading, there is no horizontal deflexion of the load point, O.

From equation (i) of example 2.12 we may derive an expression for strain energy in the frame which in this case consists of non-linear elastic members having an initial lack of fit $\lambda$. Since the total strain is given by

$$\varepsilon_T = \dfrac{\Delta_T - \lambda}{L}$$

we have

$$U = \sum_1^3 \dfrac{3A}{4}\left(\dfrac{B}{L}\right)^{1/3} (\Delta_T - \lambda)^{4/3} \qquad \text{(i)}$$

Applying Castigliano's first theorem, part I, and noting that $\lambda$ is independent of $\Delta_H$ and $\Delta_V$, we have

$$\dfrac{\partial U}{\partial \Delta_H} = \sum_1^3 A\left[\dfrac{B}{L}(\Delta_T - \lambda)\right]^{1/3} \dfrac{\partial \Delta_T}{\partial \Delta_H} = 0$$

and $\dfrac{\partial U}{\partial \Delta_V} = \sum_1^3 A\left[\dfrac{B}{L}(\Delta_T - \lambda)\right]^{1/3} \dfrac{\partial \Delta_T}{\partial \Delta_V} = 10 \text{ kN}$

Since A, B and L are the same for all members

$$\sum_{1}^{3} (\Delta_T - \lambda)^{1/3} \frac{\partial \Delta_T}{\partial \Delta_H} = 0 \qquad \text{(ii)}$$

and $\sum_{1}^{3} (\Delta_T - \lambda)^{1/3} \frac{\partial \Delta_T}{\partial \Delta_V} = \frac{10}{A}(\frac{L}{B})^{1/3}$ (iii)

From table 2.15 we obtain the terms to be substituted in these equations (refer to table 2.12). Notice that although $\Delta_H$ is eventually to be set to zero, we still need to take account of it here (as a dummy deflexion) in order to evaluate $\partial \Delta_T / \partial \Delta_H$.

Table 2.15

| Member | $(\Delta_T - \lambda)$ | $\frac{\partial \Delta_T}{\partial \Delta_H}$ | $\frac{\partial \Delta_T}{\partial \Delta_V}$ |
|---|---|---|---|
| AO | $\frac{\Delta_V + \sqrt{3}\Delta_H}{2}$ | $\frac{\sqrt{3}}{2}$ | $\frac{1}{2}$ |
| BO | $\Delta_V$ | 0 | 1 |
| CO | $\frac{\Delta_V - \Delta_H}{\sqrt{2}} - \lambda$ | $-\frac{1}{\sqrt{2}}$ | $\frac{1}{\sqrt{2}}$ |

Thus in equations (ii) and (iii) and setting $\Delta_H$ equal to zero, we have

$$\left(\frac{\Delta_V}{2}\right)^{1/3} \frac{\sqrt{3}}{2} - \left(\frac{\Delta_V}{\sqrt{2}} - \lambda\right)^{1/3} \frac{1}{\sqrt{2}} = 0 \qquad \text{(iv)}$$

and $\left(\frac{\Delta_V}{2}\right)^{1/3} \frac{1}{2} + \Delta_V^{1/3} + \left(\frac{\Delta_V}{2} - \lambda\right)^{1/3} \frac{1}{\sqrt{2}} = \frac{10}{A}\left(\frac{L}{B}\right)^{1/3}$ (v)

From equation (iv) we obtain

$$\Delta_V = \frac{4\sqrt{2}\lambda}{4 - 3\sqrt{3}}$$

Substituting for $\Delta_V$ in equation (v) we have

$$\left(\frac{\lambda}{4 - 3\sqrt{3}}\right)^{1/3} \left(\frac{1 + \sqrt{3} + 2^{4/3}}{\sqrt{2}}\right) = \frac{10}{A}\left(\frac{L}{B}\right)^{1/3}$$

49

from which

$$\lambda = -\frac{1000L}{A^3B} \frac{2\sqrt{2}(3\sqrt{3} - 4)}{(1 + \sqrt{3} + 2^{4/3})^3}$$

or $\lambda = -23.35 \dfrac{L}{A^3B}$

thus $\Delta_V = 110.45 \dfrac{L}{A^3B}$

To determine the member forces we note that

$$F_T = A\left(\frac{B}{L}\right)^{1/3} (\Delta_T - \lambda)^{1/3}$$

thus $(F_T)_{OA} = \left(\dfrac{110.45}{2}\right)^{1/3} = 3.81$ kN

$(F_T)_{OB} = (110.45)^{1/3} = 4.80$ kN

and $(F_T)_{OC} = \left(\dfrac{110.45}{\sqrt{2}} + 23.35\right)^{1/3} = 4.66$ kN

These member forces should be compared with those found in examples 2.2 and 2.12 where no lack of fit was involved. The negative sign found for $\lambda$ indicates that the unstressed length of member CO is *shorter* than L by this amount.

## 2.6 DEFLEXIONS IN PIN-JOINTED FRAMES

The first theorem of complementary energy (equation 1.2) may be used to determine deflexions in any type of structure. The complementary energy is a function of the loads and it must be remembered that the partial differential coefficient of the complementary energy with respect to the load $P_j$ gives the deflexion at, and in the direction of, the load $P_j$. It is often necessary to determine a deflexion at some point where no external load exists: in this case, a fictitious, or dummy load is introduced in order to obtain the partial differential coefficient. Thereafter, the dummy load is set to zero.

With one exception, it is possible to make direct use of strain energy for the special case of a single external load (W) acting on a statically determinate or indeterminate structure where the only deflexion ($\Delta$) required is that of the load itself. The strain energy

(U) stored is then equal to the external work done, thus, for linear elastic materials

$$U = \tfrac{1}{2}W\Delta$$

or $\Delta = \dfrac{2U}{W}$

The exceptional circumstance arises in statically indeterminate structures due to the presence of self-straining brought about by the initial lack of fit of the members or the effect of temperature changes.

Although the direct use of strain energy is a trivial instance of the application of energy methods to the determination of deflexions it may occasionally be of value in the solution of certain problems.

In any structure, a complete force analysis is normally the first step to determining deflexions. An exception is where the equilibrium method (section 2.5) is used. This gives deflexions directly but is limited in its application since, in practice, most structures have many more degrees of freedom than redundant forces. This is certainly so in the case of pin-jointed frames with which this section is concerned.

To illustrate the process of obtaining deflexions using the dummy load we consider the following elementary example.

*Example 2.15*

The plane pin-jointed cantilever frame ABC shown in figure 2.22a is subjected to a vertical load of 10 kN at C. Determine the vertical and horizontal deflexions at C if the bar material is linear elastic and EA is 10 MN.

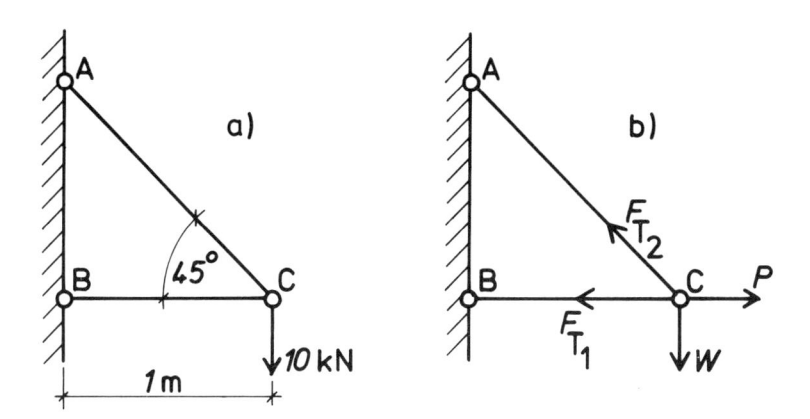

Figure 2.22

Since we require both displacements at C it is necessary to ensure that external forces are present at this point acting in the direction of the desired displacements. These forces are shown in figure 2.22b where W = 10 kN and P = 0. It is essential to deal with these forces in symbolic form since only then is it possible to identify the partial differential coefficients of the complementary strain energy.

From figure 2.22b we obtain the values of the member forces as

$$F_{T_1} = P - W \qquad \text{(i)}$$

$$F_{T_2} = \sqrt{(2)}W \qquad \text{(ii)}$$

The complementary strain energy in the frame is given by

$$C = \sum_1^2 \frac{F_T^2 L}{2EA}$$

and from the first theorem of complementary strain energy (equation 1.2) we have

$$\frac{\partial C}{\partial P} = \sum_1^2 \frac{F_T L}{EA} \frac{\partial F_T}{\partial P} = \Delta_H \qquad \text{(iii)}$$

and $$\frac{\partial C}{\partial W} = \sum_1^2 \frac{F_T L}{EA} \frac{\partial F_T}{\partial W} = \Delta_V \qquad \text{(iv)}$$

From equations (i) and (ii)

$$\frac{\partial F_{T_1}}{\partial P} = 1, \quad \frac{\partial F_{T_2}}{\partial P} = 0$$

$$\frac{\partial F_{T_1}}{\partial W} = -1, \quad \frac{\partial F_{T_2}}{\partial W} = \sqrt{2}$$

Substituting these values in equations (iii) and (iv) and noting that $L_1 = 1$ m, $L_2 = \sqrt{2}$ m, we obtain

$$EA\Delta_H = (P - W) \times 1 \times 1 + \sqrt{(2)}W \times \sqrt{2} \times 0$$

and $$EA\Delta_V = (P - W) \times 1(-1) + \sqrt{(2)}W \times \sqrt{2} \times \sqrt{2}$$

but P = 0, W = 10 kN and EA = $10^4$ kN, thus

$$\Delta_H = \frac{-10}{10^4} \text{ m} = -1 \text{ mm}$$

and $\Delta_V = \dfrac{10 + 20\sqrt{2}}{10^4}$ m = 3.83 mm

If the vertical deflexion alone is required, the direct strain energy approach could be used, for then

$$\Delta_V = \dfrac{2U}{W} = \dfrac{2}{W} \sum_1^2 \dfrac{F_T^2 L}{2EA}$$

or $\Delta_V = \dfrac{1}{10} \left( \dfrac{100 \times 1 + 2 \times 100\sqrt{2}}{10^4} \right)$ m

$= \left( \dfrac{1 + 2\sqrt{2}}{10^3} \right)$ m = 3.83 mm

The negative sign for the horizontal deflexion indicates that the movement of C is in a direction opposite to that assumed for P.

Settlement of supports can be treated as a deflexion problem as the next example shows-

*Example 2.16*

The pin-jointed structure shown in figure 2.23 represents a swing bridge carrying a 10 kN load at H. It is supported on rollers at A and G and a hinge at J.

If the support at J settles 6 mm, calculate the reaction at J due to the load and settlement together. The structure is made of linear elastic bars for which EA = 10 MN.          (Strathclyde)

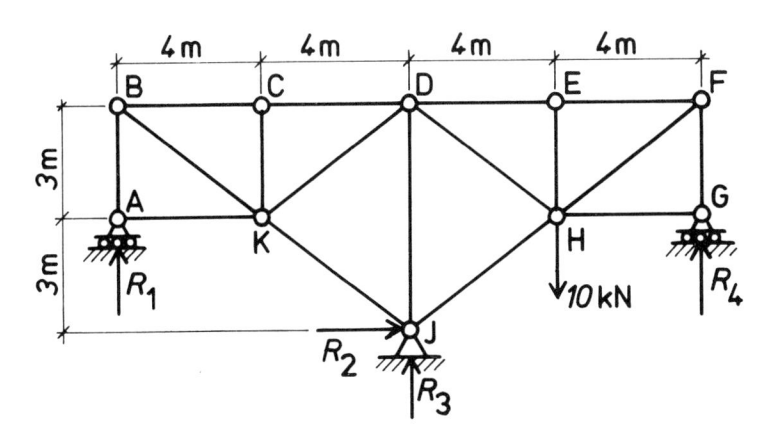

Figure 2.23

This structure has 17 members, 10 joints and there are 4 independent reactions. The structure is thus statically indeterminate with one redundant force which we will take to be the vertical reaction $R_3$ at J.

By resolving forces and taking moments about A we have

$$R_1 = \frac{5 - R_3}{2}$$

$$R_2 = 0$$

$$R_4 = \frac{15 - R_3}{2}$$

If $F_T$ represents the total axial force in each member, the complementary strain energy for the whole frame is

$$C = \sum_1^{17} \frac{F_T^2 L}{2EA}$$

Applying the first theorem of complementary energy for the vertical settlement at J, we have

$$\frac{\partial C}{\partial R_3} = \sum_1^{17} \frac{F_T L}{EA} \frac{\partial F_T}{\partial R_3} = \Delta_J \qquad (i)$$

Since the settlement at J is opposite in direction to that assumed for the reaction $R_3$

$$\Delta_J = -0.006 \text{ m}$$

also EA = 10 000 kN, thus equation (i) becomes

$$\sum_1^{17} F_T L \frac{\partial F_T}{\partial R_3} = -60 \text{ kN m} \qquad (ii)$$

The force analysis for the frame may now be carried out in the normal way giving $F_T$ in terms of $R_3$. The results are given in table 2.16 together with other relevant data.

Table 2.16

| Member | $F_T$ | L (m) | $\partial F_T/\partial R_3$ | $F_T L \dfrac{\partial F_T}{\partial R_3}$ |
|---|---|---|---|---|
| AB | $(R_3-5)/2$ | 3 | 1/2 | $3(R_3-5)/4$ |
| BK | $5(5-R_3)/6$ | 5 | $-5/6$ | $125(R_3-5)/36$ |
| BC | $2(R_3-5)/3$ | 4 | 2/3 | $16(R_3-5)/9$ |
| AK | 0 | 4 | 0 | 0 |
| CK | 0 | 3 | 0 | 0 |
| CD | $2(R_3-5)/3$ | 4 | 2/3 | $16(R_3-5)/9$ |
| DK | 0 | 5 | 0 | 0 |
| KJ | $5(5-R_3)/6$ | 5 | $-5/6$ | $125(R_3-5)/36$ |
| DJ | $-5$ | 6 | 0 | 0 |
| DE | $2(R_3-15)/3$ | 4 | 2/3 | $16(R_3-15)/9$ |
| DH | $25/3$ | 5 | 0 | 0 |
| HJ | $5(5-R_3)/6$ | 5 | $-5/6$ | $125(R_3-5)/36$ |
| HE | 0 | 3 | 0 | 0 |
| EF | $2(R_3-15)/3$ | 4 | 2/3 | $16(R_3-15)/9$ |
| FH | $5(15-R_3)6$ | 5 | $-5/6$ | $125(R_3-15)/36$ |
| HG | 0 | 4 | 0 | 0 |
| FG | $(R_3-15)/2$ | 3 | 1/2 | $3(R_3-15)/4$ |
|  |  | $\Sigma =$ |  | $\dfrac{405R_3-3425}{18}$ |

After substitution in equation (ii) we have

$R_3 = 5.8$ kN

This result may be compared with the value of $R_3$ which would have been obtained for no settlement, for then the right-hand side of equation (ii) is zero and

$R_3 = \dfrac{3425}{405} = 8.46$ kN

Similarly, the settlement required to make $R_3$ zero is obtained from equation (ii) as

$\Delta_J = \dfrac{3425}{18} \times 10^{-4}$ m = 19 mm

A final example in this section will illustrate the procedure for obtaining deflexions in statically indeterminate frames of non-linear elastic material.

*Example 2.17*

The cantilever frame shown in figure 2.24 carries loads of 10 kN at C and 20 kN at B. The bars are made from a material having a relationship between stress ($\sigma$) and strain ($\varepsilon$) given by

$$\sigma^3 = B\varepsilon$$

The cross-sectional area of each bar is A and $BA^3 = 7$ GN.

Determine the wall reactions at A, D and F and the vertical deflexion at joint C.

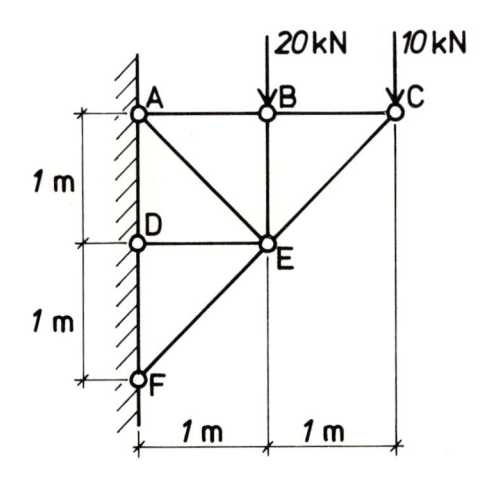

Figure 2.24

The cantilever frame has 7 members, 6 joints and there are 6 independent reactions; it is therefore statically indeterminate with one redundancy. We shall take this redundancy to be the force in member AE and proceed to examine the three structural systems shown in figure 2.25.

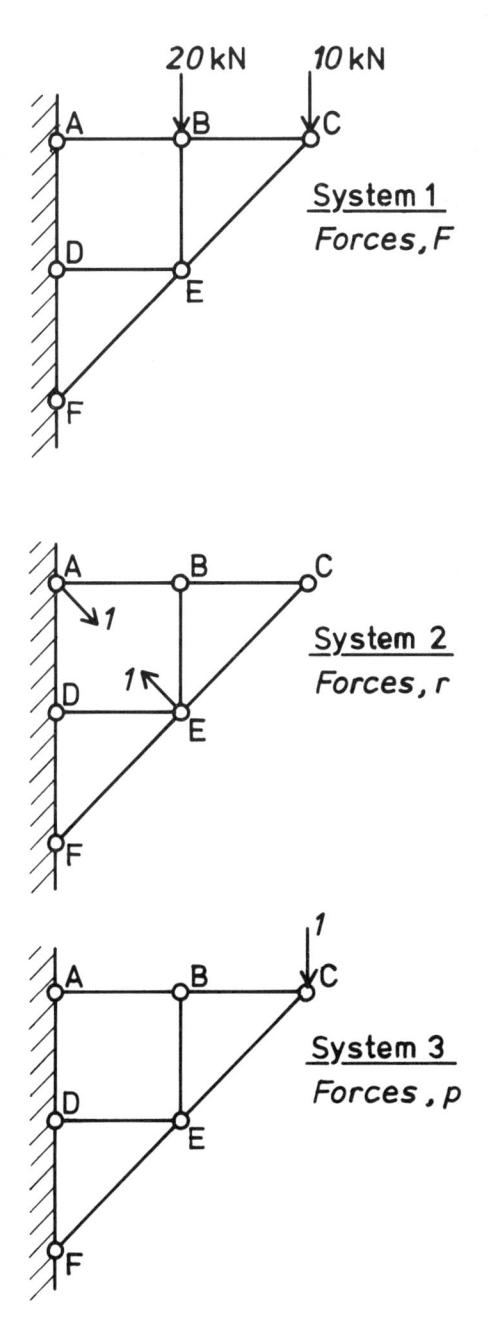

Figure 2.25

From figure 2.25, the total force in each member is

$$F_T = F + rR + pP$$

where R is the force in member AE and P is a dummy vertical load at C.

By referring back to example 2.2, we obtain the complementary strain energy for an assemblage of bars having the non-linear elastic relationship given, as

$$C = \sum_1^7 \frac{F_T^4 L}{4BA^3}$$

Making use of the first and second theorems of complementary energy (equations 1.2 and 1.6) we have

$$\frac{\partial C}{\partial P} = \sum_1^7 \frac{F_T^3 L}{BA^3} \frac{\partial F_T}{\partial P} = \Delta_C \qquad (i)$$

and $\quad \dfrac{\partial C}{\partial R} = \sum_1^7 \dfrac{F_T^3 L}{BA^3} \dfrac{\partial F_T}{\partial R} = 0 \qquad (ii)$

hence

$$\sum_1^7 (F + rR)^3 \, pL = BA^3 \Delta_C \qquad (iii)$$

and $\quad \sum_1^7 (F + rR)^3 \, rL = 0 \qquad (iv)$

since the dummy load, P is zero.

Expanding equations (iii) and (iv) we obtain

$$\sum_1^7 F^3 pL + 3R\sum_1^7 F^2 rpL + 3R^2\sum_1^7 Fr^2 pL + R^3\sum_1^7 r^3 pL = BA^3\Delta_C \qquad (v)$$

and $\quad \sum_1^7 F^3 rL + 3R\sum_1^7 F^2 r^2 L + 3R^2\sum_1^7 Fr^3 L + R^3\sum_1^7 r^4 L = 0 \qquad (vi)$

Table 2.17 lists values of F, r, and p obtained from the force analyses of the three structural systems shown in figure 2.25.

Table 2.17

| Member | L (m) | F (kN) | r | p |
|---|---|---|---|---|
| AB | 1 | +10 | 0 | 1 |
| BC | 1 | +10 | 0 | 1 |
| AE | √2 | 0 | 1 | 0 |
| BE | 1 | -20 | 0 | 0 |
| CE | √2 | -10√2 | 0 | -√2 |
| DE | 1 | +20 | -√2 | 0 |
| EF | √2 | -30√2 | 1 | -√2 |

From the table we obtain the following summations needed for equations (v) and (vi)

$$\sum_1^7 F^3 pL = 2000\,(1 + 56\sqrt{2}) = 160.4 \times 10^3 \text{ kN}^3 \text{ m}$$

$$\sum_1^7 F^2 rpL = -3600 \text{ kN}^2 \text{ m}$$

$$\sum_1^7 F r^2 pL = 60\sqrt{2} = 84.85 \text{ kN m}$$

$$\sum_1^7 r^3 pL = -2 \text{ m}$$

$$\sum_1^7 F^3 rL = -4000\,(27 + 2\sqrt{2}) = -119.3 \times 10^3 \text{ kN}^3 \text{ m}$$

$$\sum_1^7 F^2 r^2 L = 200\,(4 + 9\sqrt{2}) = 3345.6 \text{ kN}^2 \text{ m}$$

$$\sum_1^7 F r^3 L = -20\,(3 + 2\sqrt{2}) = -116.6 \text{ kN m}$$

$$\sum_1^7 r^4 L = 2\,(2 + \sqrt{2}) = 6.83 \text{ m}$$

Substitution of the summations into equations (v) and (vi) gives

$$\frac{BA^3 \Delta_C}{2000} = 80.2 - 54 \left(\frac{R}{10}\right) + 12.7 \left(\frac{R}{10}\right)^2 - \left(\frac{R}{10}\right)^3 \qquad \text{(vii)}$$

and $\left(\frac{R}{10}\right)^3 - 5.1 \left(\frac{R}{10}\right)^2 + 14.7 \left(\frac{R}{10}\right) - 17.47 = 0$ \qquad (viii)

Equation (viii) has only one real root and, by trial, this is found to be 2.074, thus

$$R = 20.74 \text{ kN}$$

After substituting for R and $BA^3$ in equation (vii) we obtain

$$\Delta_C = 4 \text{ mm}$$

The final member forces are given below. The wall reactions derived from these forces are shown in figure 2.26.

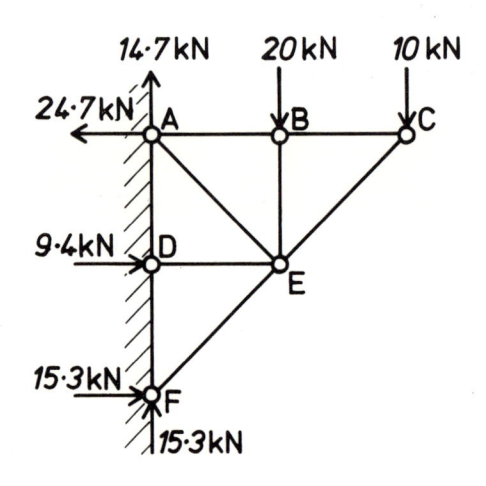

$(F_T)_{AB} = +10.0$ kN

$(F_T)_{BC} = +10.0$ kN

$(F_T)_{AE} = +20.7$ kN

$(F_T)_{BE} = -20.0$ kN

$(F_T)_{CE} = -14.1$ kN

$(F_T)_{DE} = -9.4$ kN

$(F_T)_{EF} = -21.7$ kN

Figure 2.26

## 2.7 DESIGN EXAMPLE

Although this book is concerned with analysis rather than design, it is useful to examine a practical problem which draws together the various concepts covered in this chapter. The problem is, of course, much more complex than would be expected for an examination question. The design element is introduced by requiring a search for the most economical section.

The example concerns the design of a mounting for a special testing machine. The mounting is in the form of the pin-jointed plane frame ABCDE shown in figure 2.27. The support at D offers restraint in both horizontal and vertical directions whilst the support at A is free to slide in the vertical direction. The forces transmitted to the mounting consist of 5 kN vertically and 10 kN horizontally at both B and C.

The mounting is to be constructed, as economically as possible, using aluminium alloy tube of the same section throughout. The testing machine and its mounting are normally operated in a temperature controlled environment, but the mounting must be designed to accommodate a temperature drop of up to 25 K in members AB, BC and CD. Under no circumstances is the vertical deflexion at joint B to exceed 1.2 mm.

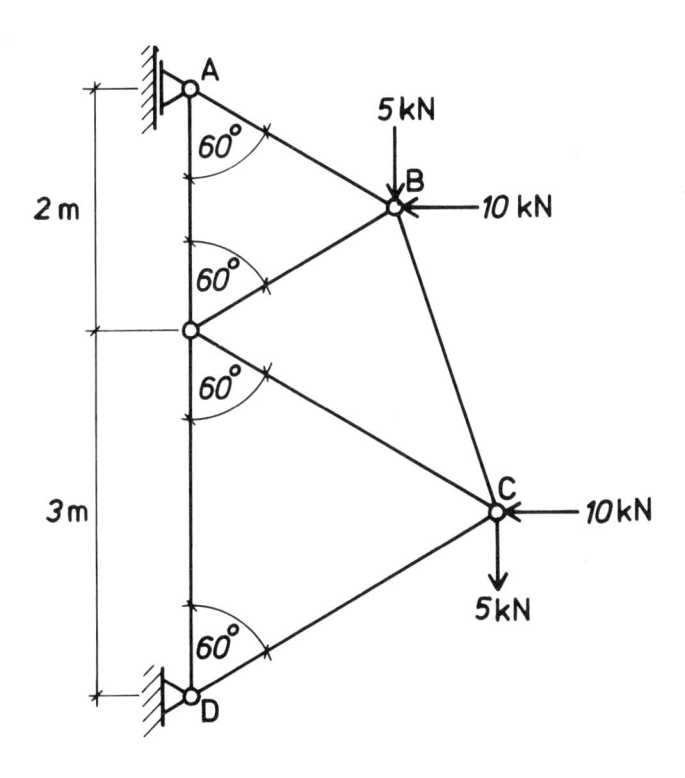

Figure 2.27

As a first step, the member forces (F) in the statically determinate mounting of figure 2.27 are obtained in the usual way. The values are shown in table 2.19 and compared with corresponding maximum allowable loads selected from the safe load table.

Table 2.18 shows the safe loads which may be carried by aluminium alloy tubes of diameter D and wall thickness t.

Table 2.18

| D (mm) | t (mm) | A (mm²) | Max. axial tension (kN) | \multicolumn{5}{c}{Max. axial compression in kN for effective length in m} |
|---|---|---|---|---|---|---|---|---|
| | | | | 2.0 | 2.5 | 3.0 | 4.0 | 4.5 |
| 42 | 3 | 394 | 51 | 10 | 6.5 | 4.5 | 2.5 | 2.0 |
| | 4 | 483 | 63 | 11 | 7.5 | 5.5 | 3.0 | 2.5 |
| 48 | 3 | 453 | 59 | 15 | 9.5 | 7.0 | 3.8 | 3.0 |
| | 4 | 557 | 72 | 17 | 11 | 8.0 | 4.5 | 3.6 |
| | 5 | 680 | 88 | 21 | 13 | 9.5 | 5.5 | 4.2 |
| 60 | 4 | 707 | 92 | 35 | 23 | 16 | 9.5 | 7.5 |
| | 5 | 869 | 113 | 42 | 27 | 19 | 11 | 8.5 |
| 76 | 4 | 906 | 118 | 68 | 47 | 33 | 19 | 15 |
| | 5 | 1120 | 146 | 82 | 56 | 40 | 23 | 18 |

For aluminium alloy, $E = 70$ GN m$^{-2}$, $\alpha = 23 \times 10^{-6}$ K$^{-1}$

Table 2.19

| Member | F (kN) | L (mm) | \multicolumn{3}{c}{Max. allowable load (kN)} |
|---|---|---|---|---|---|
| | | | D=48 mm t=5 mm | D=60 mm t=4 mm | D=60 mm t=5 mm |
| AB | − 7.70 | 2 | −21 | −35 | −42 |
| BC | − 6.11 | 2.65 | −12 | −21 | − 24 |
| CD | −15.40 | 3 | − 9.5 | −16 | − 19 |
| DE | − 2.30 | 3 | − 9.5 | −16 | − 19 |
| AE | + 3.85 | 2 | +88 | +92 | +113 |
| BE | − 6.15 | 2 | −21 | −35 | − 42 |
| CE | + 6.15 | 3 | +88 | +92 | +113 |

The negative sign denotes compression

It is evident that the most heavily loaded member is CD, and from table 2.19 the smallest section that is adequate has D = 60 mm and t = 4 mm.

The second step is to check the vertical deflexion at B, not only under the action of the applied loads but also due to the temperature change ($\theta$) in members AB, BC and CD. To do this, we remove the applied loads and place a dummy vertical load P at B. For convenience, we determine the member forces p due to a unit value of the dummy load.

The total member force is then given by

$$F_T = F + pP$$

62

The complementary strain energy in the frame is now

$$C = \sum_1^7 \left( \frac{F_T^2 L}{2EA} + F_T \lambda \right)$$

where $\lambda = L\alpha\theta$

From the first theorem of complementary strain energy (equation 1.2) we have

$$\frac{\partial C}{\partial P} = \sum_1^7 \left( \frac{F_T L}{EA} + \lambda \right) \frac{\partial F_T}{\partial P} = \Delta_B$$

where $\partial F_T/\partial P = p$, thus

$$\Delta_B = \sum_1^7 \left[ (F + pP) \frac{L}{EA} + L\alpha\theta \right] p$$

Since P is a dummy load and is therefore zero, we have

$$\Delta_B = \frac{1}{EA} \sum_1^7 FpL + \sum_1^7 pL\alpha\theta$$

Table 2.20 lists the values of F, p and L for each of the members and gives the product terms necessary for calculating $\Delta_B$. Note that $\alpha = 23 \times 10^{-6}$ K$^{-1}$.

Table 2.20

| Member | F (kN) | p | L (m) | FpL (kN m) | pLαθ (m) |
|---|---|---|---|---|---|
| AB | − 7.7 | 0.4 | 2 | − 6.16 | 18.4θ × 10$^{-6}$ |
| BC | − 6.11 | −0.53 | 2.65 | + 8.55 | −32.2θ × 10$^{-6}$ |
| CD | −15.40 | −0.4 | 3 | +18.48 | −27.6θ × 10$^{-6}$ |
| DE | − 2.30 | −0.8 | 3 | + 5.52 | 0 |
| AE | + 3.85 | −0.2 | 2 | − 1.54 | 0 |
| BE | − 6.15 | −0.6 | 2 | + 7.38 | 0 |
| CE | + 6.15 | +0.6 | 3 | +11.07 | 0 |

$$\sum_1^7 = +43.3, \qquad = -41.4\theta \times 10^{-6}$$

Hence

$$\Delta_B = \frac{43.3}{EA} - 41.4\theta \times 10^{-6} \text{ m}$$

The deflexion at B is thus compounded of two factors, one which is dependent on the material and the cross-sectional area (A) and one which is dependent on the material only.

On a strength criterion alone, the section chosen for the frame had D = 60 mm, t = 4 mm and A = 707 mm$^2$. Since E is $70 \times 10^6$ kN m$^{-2}$, we have EA = 49490 kN and so

$$\Delta_B = 0.875 - 0.0414\theta \text{ mm}$$

Under normal operating conditions, $\theta = 0$ and $\Delta_B = 0.875$ mm, which is well within the specified limiting deflexion of 1.2 mm. However, when members AB, BC and CD suffer a temperature drop of 25 K we have

$$\Delta_B = 0.875 - 0.0414(-25) = 1.91 \text{ mm}$$

This deflexion is unacceptable, but one way in which the deflexion at B could be brought down to the limiting value of 1.2 mm is to increase the cross-sectional area. From the general expression for $\Delta_B$ above with $\theta = -25$ K and $\Delta_B = 1.2$ mm we have

$$\Delta_B = \frac{43.3}{EA} - 41.4(-25) \times 10^{-6} = 1.2 \times 10^{-3} \text{ m}$$

hence

$$EA = 262.5 \times 10^3 \text{ kN}$$

from which

$$A = 3750 \text{ mm}^2$$

The use of such a large section simply to control deflexions would clearly be uneconomical, thus the alternative solution of inserting an additional member between B and D will be adopted. The presence of this member makes the frame statically indeterminate and it is therefore necessary to make use of the second theorem of complementary energy (equation 1.6) in order to carry out the force analysis.

If the force in the redundant member BD is R then the total member forces due to the external loads, the dummy load P at B and the force in the redundant member are given by

$$F_T = F + pP + rR$$

where r is the member force due to a unit force in the redundant member.

For completeness we shall suppose that member BD is inserted into the frame with an initial lack of fit of a m. Then the complementary strain energy for the frame is given by

$$C = \sum_1^8 \left( \frac{F_T^2 L}{2EA} + F_T \lambda \right)$$

where $\lambda = L\alpha\theta$ for members AB, BC and CD and $\lambda = a$ for member BD. Thus

$$\frac{\partial C}{\partial R} = \sum_1^8 \left( \frac{F_T L}{EA} + \lambda \right) r = 0$$

and

$$\frac{\partial C}{\partial P} = \sum_1^8 \left( \frac{F_T L}{EA} + \lambda \right) p = \Delta_B$$

Since P = 0, these equations become

$$\sum_1^8 FrL + R\sum_1^8 r^2L + EA\sum_1^8 r\lambda = 0 \qquad \text{(i)}$$

and $\sum_1^8 FpL + R\sum_1^8 rpL + EA\sum_1^8 p\lambda = EA\Delta_B$ (ii)

Table 2.21 lists the values of r and gives the additional product terms required for equations (i) and (ii).

Table 2.21

| Member | r | FrL (kN m) | $r^2L$ (m) | $\lambda \times 10^6$ (m) | $r\lambda \times 10^6$ (m) | rpL (m) | $p\lambda \times$ (m) |
|---|---|---|---|---|---|---|---|
| AB | 0 | 0 | 0 | +46θ | 0 | 0 | +18.4 |
| BC | -0.61 | + 9.81 | +0.975 | +60.8θ | -37θ | +0.855 | -32.2 |
| CD | -0.46 | +21.19 | +0.632 | +69θ | -31.7θ | +0.552 | -27.6 |
| DE | -0.69 | + 4.76 | +1.421 | 0 | 0 | +1.656 | 0 |
| AE | 0 | 0 | 0 | 0 | 0 | 0 | 0 |
| BE | -0.69 | + 8.47 | +0.947 | 0 | 0 | +0.828 | 0 |
| CE | +0.69 | +12.71 | +1.421 | 0 | 0 | +1.242 | 0 |
| BD | 1 | 0 | +4.359 | $a/10^6$ | $a/10^6$ | 0 | 0 |
| | Σ = | +56.94 | +9.755 | | $a/10^6$ −68.7θ | +5.133 | −41.4 |

and $\Sigma FpL = +43.3$ obtained previously

After substitution of the product terms into equations (i) and (ii) we obtain

$$56.94 + 9.755R + EA(a - 68.7\theta \times 10^{-6}) = 0$$

and $43.3 + 5.133R - EA \times 41.4\theta \times 10^{-6} = EA\Delta_B$

As a first trial solution of these equations, we shall choose the original section (D = 60 mm, t = 4 mm, A = 707 mm$^2$), then

$$5.84 + R + 5073.3a - 0.348\theta = 0 \qquad \text{(iii)}$$

and $8.43 + R - 0.40\theta = 9641.5\Delta_B$ (iv)

After eliminating R between equations (iii) and (iv) we obtain, for a in mm

$$\Delta_B = 0.27 - 5.4\theta \times 10^{-3} - 0.526a \text{ mm}$$

from which

$$a = 0.512 - 10.25\theta \times 10^{-3} - 1.9\Delta_B \text{ mm}$$

Since $\Delta_B$ is not to exceed 1.2 mm when $\theta = -25$ K we find that a is negative and should not be greater than 1.5 mm. It is discovered later that this lack of fit produces a value of R at $\theta = 0$ K, which is just too large for the section chosen. For this reason, we take $a = -1.4$ mm, then

$\Delta_B = 1.14$ mm when $\theta = -25$ K

and $\Delta_B = 1.00$ mm when $\theta = 0$ K

Substitution for a and $\theta$ in equation (iii) gives

$R = -7.44$ kN when $\theta = -25$ K

and $R = +1.26$ kN when $\theta = 0$ K

The final member forces are given in table 2.22 together with the allowable loads for the section $D = 60$ mm, $t = 4$ mm.

Table 2.22

| Member | $F_T$ (kN) $\theta = 0$ K | $F_T$ (kN) $\theta = -25$ K | Max. allowable load (kN) for $D = 60$ mm, $t = 4$ mm |
|---|---|---|---|
| AB | − 7.70  | − 7.70  | −35 |
| BC | − 6.88  | − 1.57  | −21 |
| CD | −15.97  | −11.98  | −16 |
| DE | − 3.17  | + 2.83  | −16/+92 |
| AE | + 3.85  | + 3.85  | +92 |
| BE | − 7.02  | − 1.02  | −35 |
| CE | + 7.02  | + 1.02  | +92 |
| BD | + 1.26  | − 7.44  | +92/−8 |

Thus the mounting frame in figure 2.27 may be fabricated from aluminium alloy tube of diameter 60 mm and wall thickness 4 mm, provided an additional member is placed between B and D to act as a prop under low-temperature conditions. The prop should be inserted with a tensile preload of 15.9 kN which may be achieved by making the member initially too short by 1.4 mm.

Under these circumstances both strength and deflexion checks are satisfactory.

PROBLEMS

1. The three linearly elastic bars shown in figure 2.28 are pinned together at O and to rigid supports at A, B and C. Determine the forces in each of the bars, when there is no initial lack of fit.

[$F_{OA} = F_{OC} = +29.3$ kN, $F_{OB} = +58.6$ kN]

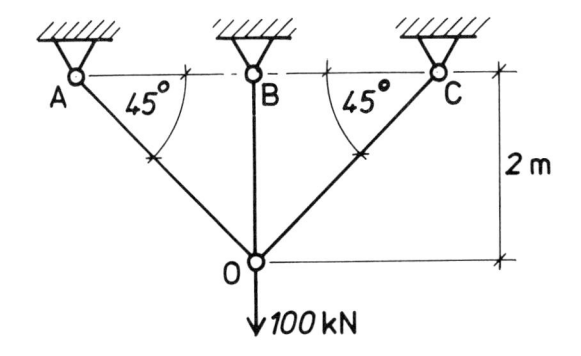

Figure 2.28

2. The frame ABCD shown in figure 2.29 is subjected to a horizontal load of 10 kN at C. Which member is carrying the greater force? What is the magnitude and sense of this force? If the roller support at D were replaced by a pinned support, what would be the new force in BD? Assume linear elastic behaviour.  [AC, +8.54 kN, -6.26 kN]

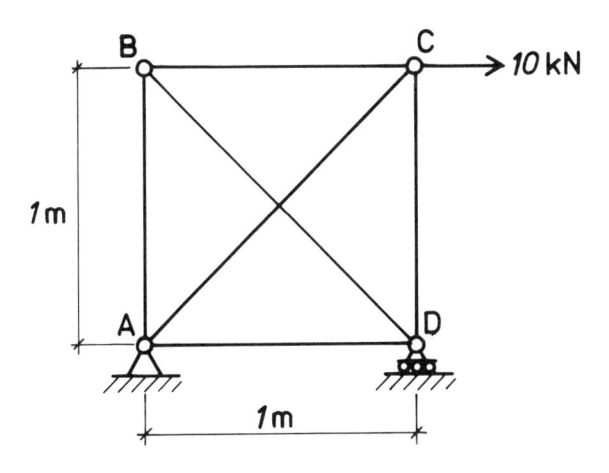

Figure 2.29

3. The pin-jointed plane frame shown in figure 2.30 is statically determinate. It is found, however, that due to unforeseen service loading, additional bracing is required between A and C and C and F to reduce the frame deflexion. Compare the forces in member CD before and after bracing. The members are made from a linearly elastic material for which EA is constant.  [+10 kN, +18 kN]

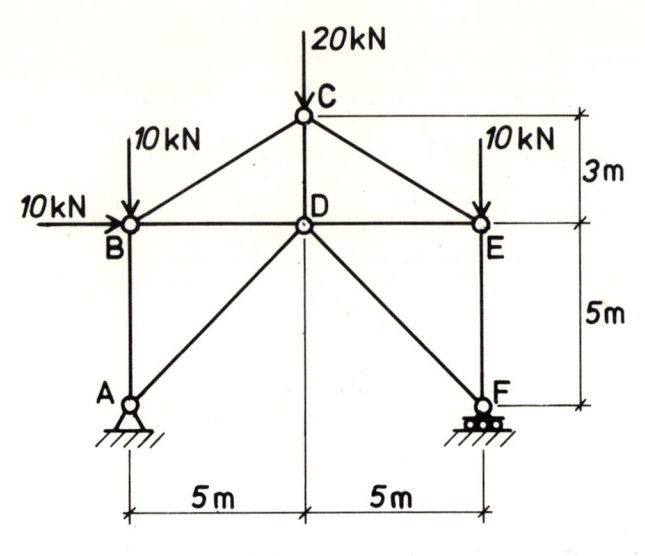

Figure 2.30

4. Determine the maximum values of tensile and compressive force in the plane pin-jointed frame shown in figure 2.31. Assume linear elastic behaviour. [+9.75 kN, -20.84 kN]

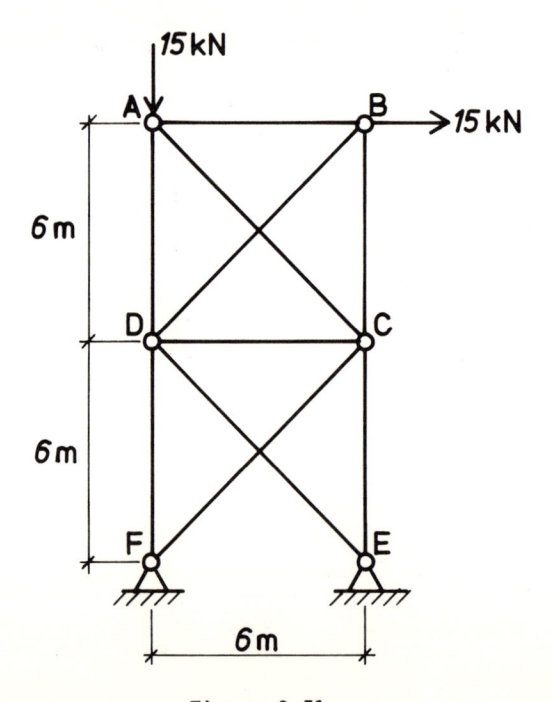

Figure 2.31

5. The member DE of the pin-jointed frame shown in figure 2.32 had to be compressed by 2.5 mm in order to be fitted into the frame. Find the force in member DE caused by this initial lack of fit and by the application of a 30 kN load at C. The cross-sectional area and modulus of elasticity of all the members are 775 mm$^2$ and 200 GN m$^{-2}$ respectively. [+7.6 kN]
[Leeds]

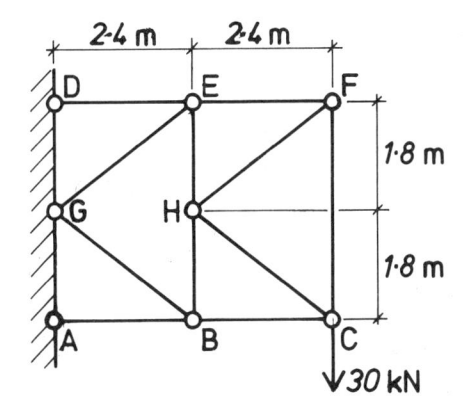

Figure 2.32

6. Each member of the truss shown in figure 2.33 has a cross-sectional area of 500 mm$^2$ and a Young's modulus of 200 GN m$^{-2}$. The turnbuckle T on diagonal member AC is tightened so that the ends of the two bars at the turnbuckle are brought 2 mm closer to each other. Determine the forces induced in each member by this operation. Assume linear elastic behaviour throughout.

[$F_{AB} = F_{BC} = F_{CD} = -10.9$ kN, $F_{BD} = F_{AC} = +15.4$ kN, $F_{AD} = 0$]
[Leicester]

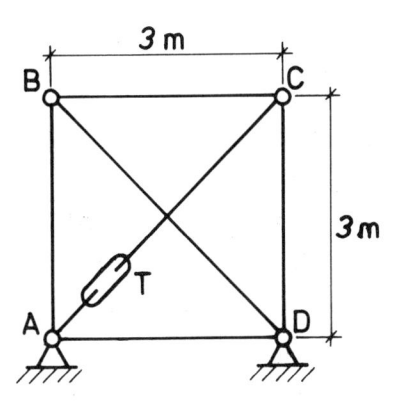

Figure 2.33

69

7. The pin-jointed linear elastic plane frame shown in figure 2.34 is pinned to rigid supports at A and B. The cross-sectional area of each member is A and the span AB is 2L. If the member DC is inserted so as to produce a thrust R at D and C, without external loading, show that the lack of fit of the member DC is

$$\frac{3RL}{2AE}(1 + \sqrt{3})$$

where E is the elastic modulus
[Leeds]

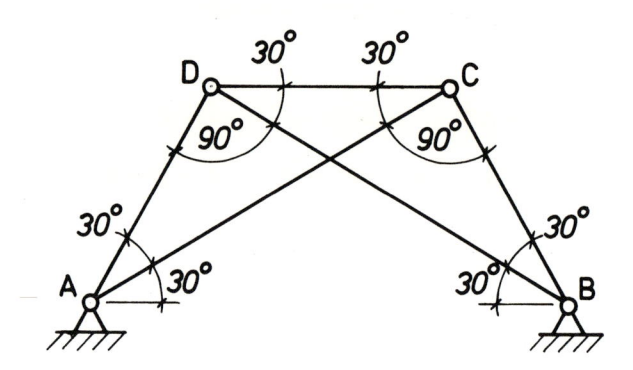

Figure 2.34

8. If the stiffness, AE/L is the same for all members of the pin-jointed frame shown in figure 2.35, show that the force in member EF due to the loads P is 9P/22. What initial lack of fit of EF would be necessary for there to be no force in AE when the loads P are applied? [-2PL/3AE]
[Leeds]

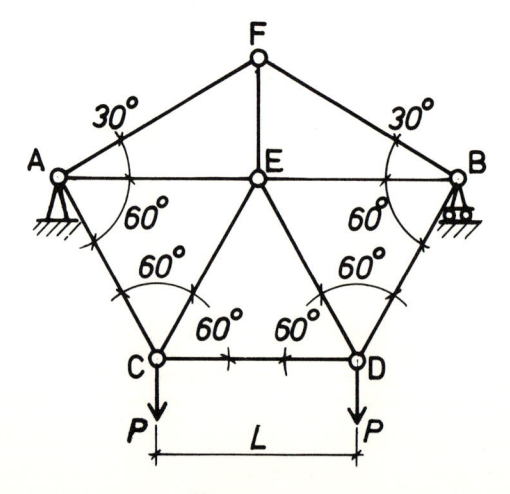

Figure 2.35

9. The pin-jointed structure shown in figure 2.36 is hinged to supports A and D and supported on a roller at E. All supports are unyielding and the support at E, although shown conventionally, is capable of providing an upward or a downward reaction.

Members BC and CF together undergo a temperature increase of 20 K relative to the temperature in the other members. For this effect alone, determine the members with the greatest force and the magnitude and nature of this force. (For all members: $AE/L = 5 \times 10^4$ kN m$^{-1}$ and $\alpha = 10^{-5}$ K$^{-1}$.
[Strathclyde]                                    [BF and CE, +14.6 kN]

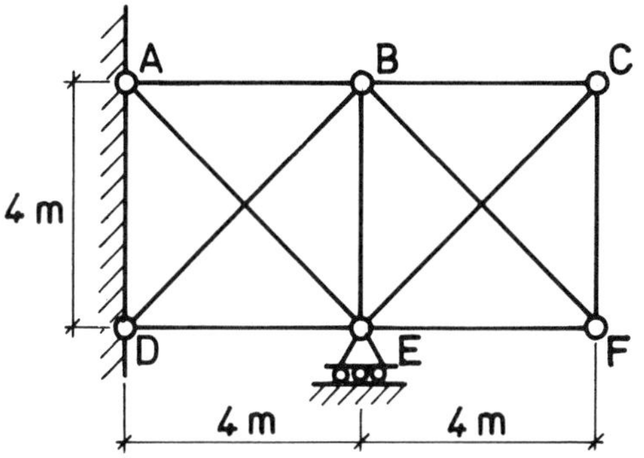

Figure 2.36

10. Member BC of the frame shown in figure 2.2 (see section 2.2) has its temperature raised by 20 K relative to the other members. Determine the total force in this member, if bars AB, CD and DE have cross-sectional areas of 200 mm$^2$ and all other bars have cross-sectional areas of 100 mm$^2$. ($E = 200$ GN m$^{-2}$ and $\alpha = 11 \times 10^{-6}$ K$^{-1}$.)
[-4.06 kN]

11. The pin-jointed steel frame shown in figure 2.37 was originally designed to have a roller support at G. On erection, however, the roller support was omitted so that both A and G are simple pin supports. The frame is part of a structure used in desert conditions where the temperature variation between day and night can be as much as 30 K. If the support G was finally tied down in the late afternoon, when the temperature was mid-way between extremes, determine the largest member force induced by this erection error. (Members AB, AC, GF and GE have cross-sectional areas of 200 mm$^2$. All other members have cross-sectional areas of 100 mm$^2$. $E = 200$ GN m$^{-2}$ and $\alpha = 11 \times 10^{-6}$ K$^{-1}$.)
[±1.12 kN in CE]

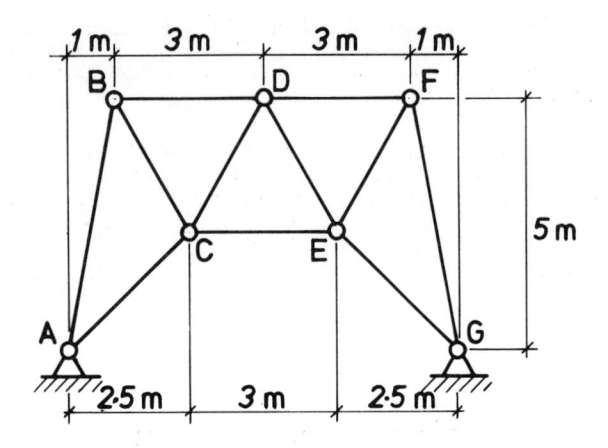

Figure 2.37

12. The pin-jointed frame ABCD shown in figure 2.38 is supported at A and D. The hinge at A is unyielding in both horizontal and vertical directions. The support at D is unyielding in the vertical direction but deflects in the horizontal direction at the rate of f m per kN of horizontal reaction. All members of the frame have the same cross-sectional area, A m² and are made from a material with a modulus of elasticity, E kN m⁻². Show that for the given loading, the horizontal reaction at D is approximately

$$\frac{7.93}{0.9+n} \text{ kN}$$

where n = AEf/L.

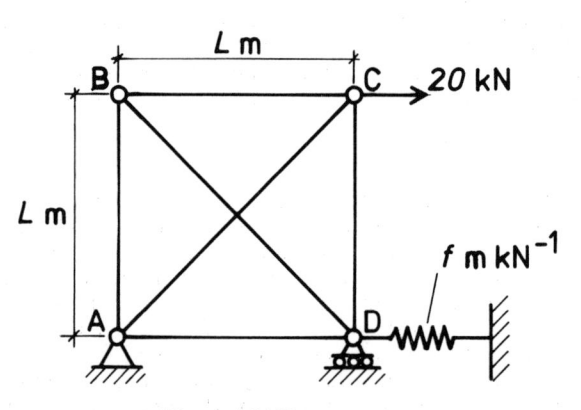

Figure 2.38

13. The frame ABC shown in figure 2.39 is supported on a hinge at A and a roller support at B. The horizontal movement (Δ) of B is

constrained by a non-linear elastic spring whose characteristic is given by

$$\Delta = 10f(0.1 + H)H$$

where H is the horizontal reaction at B in kN and f is the flexibility of member AB (L/AE).

If the members of the frame behave in a linearly elastic manner and have the same cross-sectional area and modulus of elasticity, determine the force (H) in the spring when a 6 kN load is applied vertically at C. [0.63 kN]

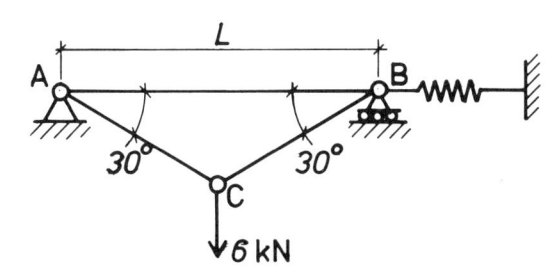

Figure 2.39

14. If the bars of the frame in problem 7 are made of a non-linear elastic material having a stress-strain relationship given by

$$\sigma^3 = B\varepsilon$$

show that the lack of fit ($\lambda$) in member DC required to produce a thrust R at D and C is given by

$$\lambda = \frac{9R^3L}{8BA^3}(1 + \sqrt{3})$$

15. Repeat problem 2 assuming that the bar material has a stress-strain relationship given by

$$\sigma^3 = B\varepsilon$$

[AC, +7.4 kN, -6.9 kN]

16. The pin-jointed plane frame ABCO shown in figure 2.40 consists of three bars of linearly elastic material and equal cross-sectional area $A_0$. The frame is subjected to an in-plane force of 50 kN applied at the joint O and inclined at 30° to the horizontal, as shown. Using the relationship

$$P_j = \frac{\partial U}{\partial \Delta_j}$$

73

where the symbols have their usual meanings, determine the horizontal and vertical displacements of the joint O and the forces in each of the three bars. Take $EA_0$ = 20 MN, where E is the modulus of elasticity. $[\Delta_V = 1.8$ mm, $\Delta_H = 1.3$ mm, $F_{AO} = 25.4$ kN,

[Sussex] $F_{BO} = 30.4$ kN, $F_{CO} = 5.0$ kN]

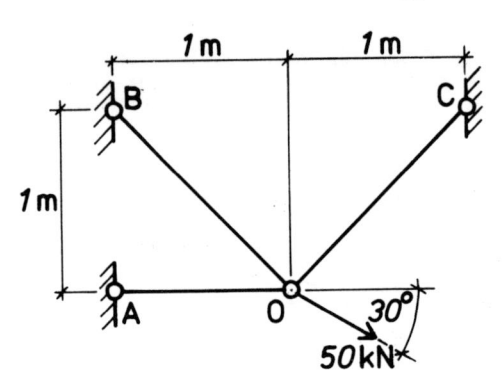

Figure 2.40

17. The pin-jointed plane frame ABCO shown in figure 2.41 consists of three straight rods all of cross-sectional area $A_0$. The frame is loaded with a *vertical* force V applied at the joint O. Members AO and BO are made of a non-linear elastic material for which the tension and compression relationship between stress ($\sigma$) and strain ($\varepsilon$) is given by

$$\sigma = 5000\, E\varepsilon^3$$

Member CO is made of a linear elastic material having a stress-strain relationship in tension and compression given by

$$\sigma = E\varepsilon$$

Using the relationship

$$P_j = \frac{\partial U}{\partial \Delta_j}$$

where the symbols have their usual meanings, determine the magnitude and sense (towards or away from O) of the applied force V if, due to this force, the horizontal deflexion of joint O is 30 mm to the right. Take $EA_0$ as 1 kN and note that strain energy (u) per unit volume due to direct stress is given by

$$u = \int \sigma\, d\varepsilon$$

[75.3 N downwards]

[Sussex]

line of action of
applied force, V

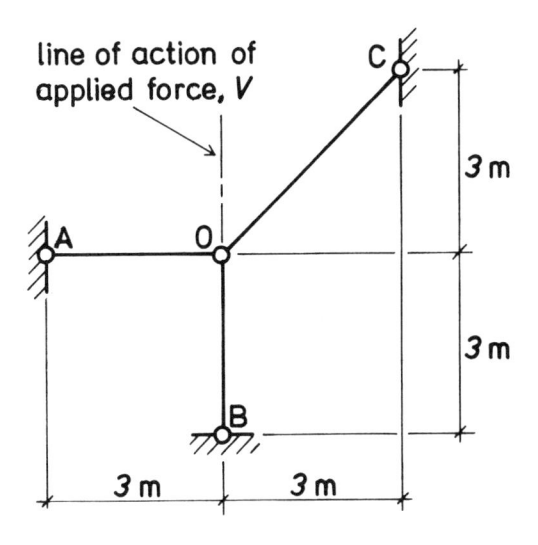

Figure 2.41

18. The pin-jointed plane frame ABCDO shown in figure 2.42 consists of four rods of linearly elastic material ($\sigma = E\varepsilon$) and equal cross-sectional area $A_0$. Determine the horizontal and vertical displacements of the load point O due to a vertical load of 10 N.
$EA_0 = 1$ kN.  [0.1 mm, 7.8 mm]
[Sussex]

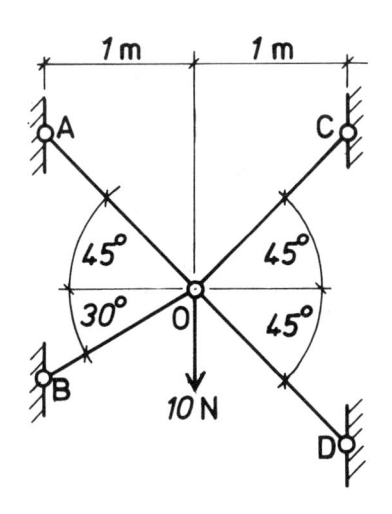

Figure 2.42

19. In the frame of problem 18, members BO and DO are replaced by rods of non-linear elastic material having the stress-strain relationship

$$\sigma = 10^4 E\varepsilon^3$$

An additional horizontal load H is applied at O of sufficient magnitude to eliminate any horizontal movement of the load point.

Determine the new vertical displacement of the load point and the value of H. [11.4 mm, 0.27 N]
[Sussex]

20. The pin-jointed plane frame ABCO shown in figure 2.43 consists of three straight rods of equal cross-sectional area and length. The frame is loaded at O with vertical and horizontal forces V and H respectively. All three members AO, BO and CO are made of a non-linear elastic material for which the relationship between tensile stress ($\sigma$) and strain ($\varepsilon$) is given by

$$\sigma = B\varepsilon^3$$

Using the relationship $P_j = \partial U/\partial \Delta_j$ where the symbols have their usual meanings, show that $V = 31H/59$ if the horizontal deflexion of O is to be twice the vertical deflexion. Note that the strain energy (u) per unit volume due to direct stress is given by $u = \int \sigma \, d\varepsilon$.
[Sussex]

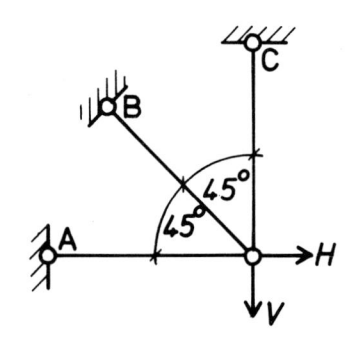

Figure 2.43

21. A load of 40 kN is suspended between two vertical walls as shown in figure 2.44 by five pin-jointed members in a vertical plane having the same axial rigidity AE = 10 MN. Calculate the horizontal and vertical displacements of the load and the force in member AF.
[London] [2.55, 13.42 mm, 9.72 kN]

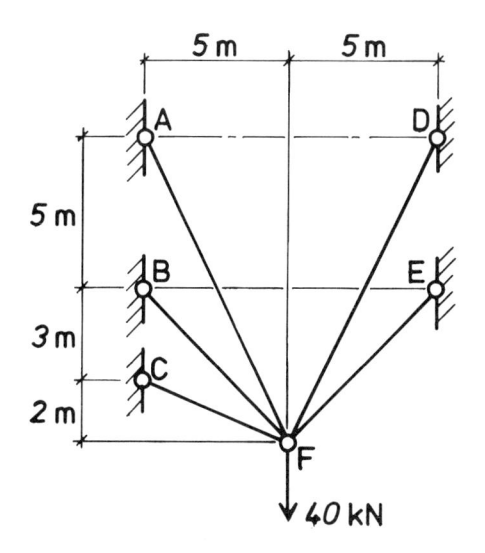

Figure 2.44

22. Two views of a pin-jointed space frame ABCDO are shown in figure 2.45. If the supports for the pin-joints at A, B, C and D are infinitely rigid, determine the components of the deflexion of the common joint O due to a uniform temperature rise of 50 K. The bar material is linear elastic and has a coefficient of thermal expansion of $12 \times 10^{-6}$ K$^{-1}$, .EA is 6000 MN.

$[\Delta_x = 0.9$ mm, $\Delta_y = -1.14$ mm, $\Delta_z = 3.0$ mm$]$

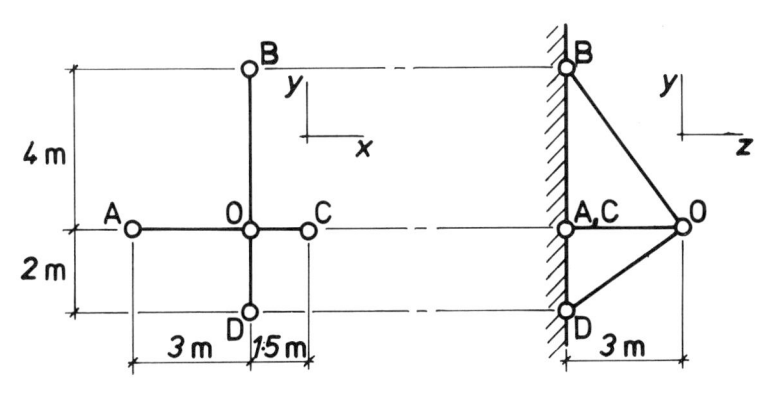

Figure 2.45

23. Show that the effect of the bracing in problem 3 is to reduce the vertical deflexion at C by approximately 16 per cent.

24. The plane truss shown in figure 2.46 is pin-jointed and all members have the same cross-sectional area, equal to 650 mm$^2$.

Determine the force in member FC when a load of 10 kN is applied at D and also the vertical component of the deflexion at this point. The bar material is linear elastic with a modulus of 200 GN m$^{-2}$.
[Sheffield] [+3.21 kN, 7.1 mm]

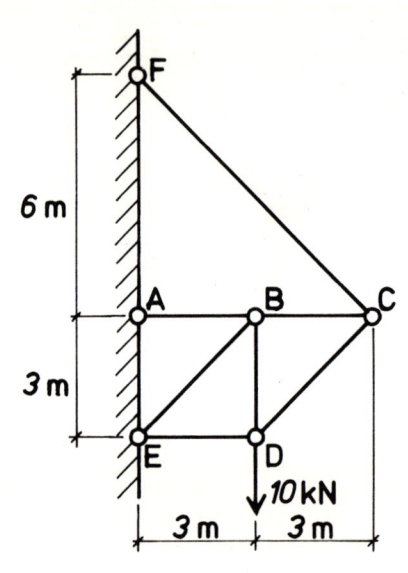

Figure 2.46

25. Calculate the vertical deflexion of the loaded point of the plane pin-jointed truss shown in figure 2.47, if the diagonals have a load-deformation relationship given by P = 44e - 8e$^2$ for e>3 mm, where P is the force in kN and e is the deformation in mm. All other members have constant axial rigidity EA = 80 MN. (Hint: Use direct strain energy.)
[London] [4.1 mm]

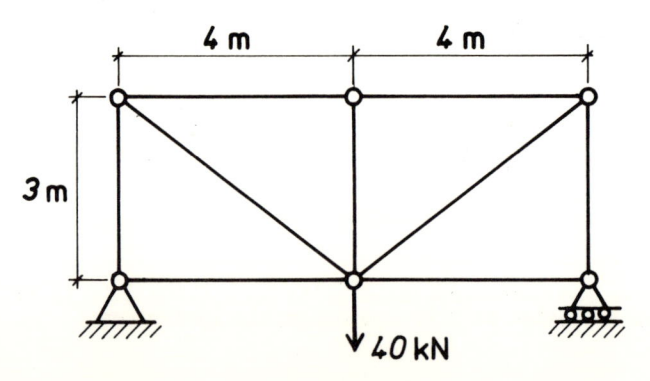

Figure 2.47

26. The Warren girder shown in figure 2.48 consists of members all having the same length and cross-sectional area. Determine the cross-sectional area so that the mid-span deflexion, when the girder is subjected to the loads shown, is not greater than 1/1000 of the span. E = 200 GN m$^{-2}$.  [1750 mm$^2$]

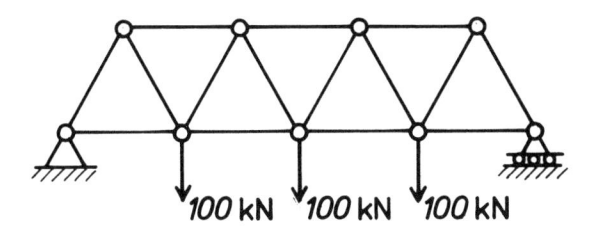

Figure 2.48

# 3 FORCE AND DEFORMATION ANALYSIS OF BEAMS, CURVED MEMBERS AND RIGID-JOINTED FRAMES

In the previous chapter we saw that, as the result of the application of external loads, only axial forces are produced in the members of pin-jointed plane and space frames. In a rigid-jointed plane frame, however, each member may be subjected to three internal forces consisting of an axial force, a shear force and a bending moment. In a rigid-jointed space frame there are six possible internal forces for each member: an axial force, two shear forces, two bending moments and one torque. The rigid-jointed space frame is outside the scope of this book, thus we shall confine our attention to plane frames subjected to in-plane loading.

It is found that unless the members of the frame are particularly short and stocky, by far the greater proportion of the work done on this type of frame is stored as bending strain energy. The contribution of axial and shear forces to the total strain energy can therefore be neglected for most conventional rigid-jointed structural systems.

Since the derivation of bending energy expressions for frames having non-linear elastic members can be extremely complex, the discussion here will be limited to frames of linear elastic material.

It should be understood that the remarks above apply equally to beams and to continuous, plane curved members.

## 3.1 COMPLEMENTARY ENERGY DUE TO BENDING

Figure 3.1 shows a small element ABCD of an initially straight member of linear elastic material subject solely to a uniform bending moment M. Since the material is linear elastic, the complementary energy stored in the element is equal to the work done by the moment as it increases from zero to its final value of M, thus

$$dC = \tfrac{1}{2} M \, d\phi \qquad \text{(i)}$$

where $d\phi$ is the rotation of face BD with respect to face AC.

From simple bending theory (see *Essential Solid Mechanics*) we have

$$\frac{1}{R} = \frac{M}{EI} \qquad \text{(ii)}$$

where EI is the flexural rigidity of the member and R is the radius of curvature.

Also, from figure 3.1, we see that

$$d\phi = \frac{dx}{R} \qquad \text{(iii)}$$

Thus from equations (i), (ii) and (iii), we have

$$dC = \frac{M^2}{2EI} dx$$

hence, if the member is of length L

$$C = \int_0^L \frac{M^2}{2EI} dx \qquad (3.1)$$

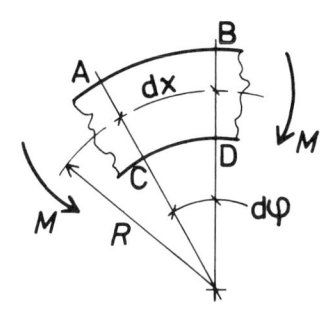

Figure 3.1

This chapter will be concerned only with the compatibility method of analysis, making use of the first and second theorems of complementary energy. The equilibrium approach is deferred until Chapter 4.

3.2 STRAIGHT BEAMS

In order to illustrate the basic procedures for dealing with members subjected to bending we shall first apply the complementary energy theorems to obtain results for some simple beam problems.

*Example 3.1*

Determine the vertical deflexion at the free end of the uniformly loaded cantilever shown in figure 3.2. The flexural rigidity of the cantilever has a constant value EI.

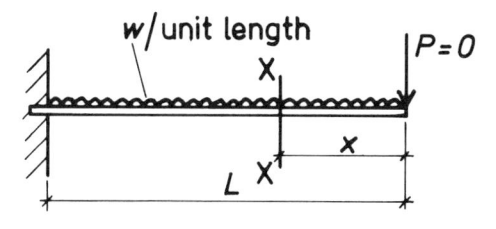

Figure 3.2

Figure 3.2 also shows the dummy vertical load P which is required at the free end of the cantilever in order to determine the vertical deflexion at this point.

The bending moment at a distance x from the free end is given by

$$M_x = \frac{wx^2}{2} + Px \qquad (i)$$

where the convention adopted is that hogging moments are positive.

The total complementary strain energy stored in the cantilever is therefore obtained from equation 3.1 as

$$C = \int_0^L \frac{M_x^2}{2EI} dx$$

Applying the first theorem of complementary energy (equation 1.2) in order to obtain the deflexion ($\Delta$) at the free end of the cantilever we have

$$\frac{\partial C}{\partial P} = \int_0^L \frac{M_x}{EI} \frac{\partial M_x}{\partial P} dx = \Delta \qquad (ii)$$

From equation (i)

$$\frac{\partial M_x}{\partial P} = + x$$

thus substituting in equation (ii) and setting the dummy load equal to zero, we obtain

$$EI\Delta = \int_0^L \frac{wx^3}{2} dx$$

from which

$$\Delta = \frac{wL^4}{8EI}$$

*Example 3.2*

Determine the slope at the free end of the cantilever of example 3.1.

Here, instead of a dummy load, P acting through a displacement $\Delta$ we require a dummy clockwise couple, M acting at the free end, through a rotation or slope, $\theta$. The moment expression for the cantilever is now

$$M_x = \frac{wx^2}{2} + M \qquad (i)$$

and 
$$\frac{\partial C}{\partial M} = \int_0^L \frac{M_x}{EI} \frac{\partial M_x}{\partial M} dx = \theta \qquad (ii)$$

From equation (i)

$$\frac{\partial M_x}{\partial M} = +1$$

thus from equation (ii) and noting that M is zero, we obtain

$$EI\theta = \int_0^L \frac{wx^2}{2} dx$$

from which

$$\theta = \frac{wL^3}{6EI}$$

*Example 3.3*

Determine the reaction R in the uniformly loaded propped cantilever AB shown in figure 3.3. The end B is propped to the same level as end A. The flexural rigidity is EI.

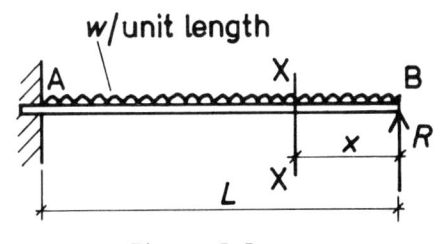

Figure 3.3

The bending moment at XX is given by

$$M_x = \frac{wx^2}{2} - Rx$$

From the expression for the complementary strain energy (equation 3.1) and the second theorem of complementary energy (equation 1.6) we have

$$\frac{\partial C}{\partial R} = \int_0^L \frac{M_x}{EI} \frac{\partial M_x}{\partial R} dx = 0 \qquad (i)$$

or  $\int_0^L \left(\frac{wx^2}{2} - Rx\right)(-x)\, dx = 0$

from which we obtain

$R = \frac{3wL}{8}$

If the prop had not been rigid but was subject to settlement $\lambda$, equation (i) becomes

$\frac{\partial C}{\partial R} = \int_0^L \frac{M_x}{EI} \frac{\partial M_x}{\partial R}\, dx = -\lambda$

since the settlement is opposite in direction to the load. Thus

$\int_0^L \left(\frac{wx^2}{2} - Rx\right)(-x)\, dx = -EI\lambda$

from which we obtain

$R = \frac{3wL}{8} - \frac{3EI\lambda}{L^3}$

Note that $R = 0$ when $\lambda$ is equal to the value of $\Delta$ obtained from example 3.1.

*Example 3.4*

Determine the reactions and the principal bending moments for the uniform continuous beam shown in figure 3.4a.

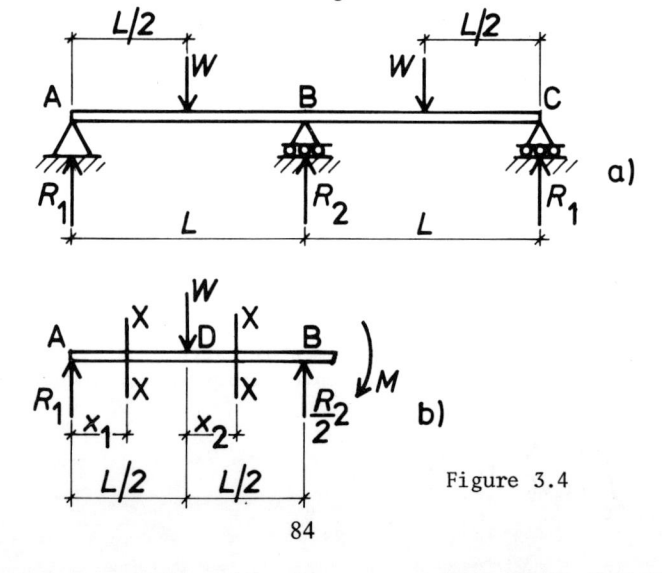

Figure 3.4

Since the beam is symmetrical about B, we need only consider half the beam, AB as shown in figure 3.4b.

Because of the nature of the supports, there is clearly no axial force at B and due to symmetry, the shear force at B is also zero. The only unknown force which remains is the moment M which is required to preserve continuity over the central support.

The bending moments in lengths AD and DB of the beam are as follows

$$M_{AD} = -R_1 x_1 \qquad 0 < x_1 < \frac{L}{2}$$

and $\quad M_{DB} = -R_1 (x_2 + \frac{L}{2}) + W x_2 \qquad 0 < x_2 < \frac{L}{2}$

Since the reaction $R_1$ may be taken as the redundant force in the system, we have, from the second theorem of complementary energy, that

$$\frac{\partial C}{\partial R_1} = 2 \int_0^{L/2} \frac{M_{AD}}{EI} \frac{\partial M_{AD}}{\partial R_1} dx_1 + 2 \int_0^{L/2} M_{DB} \frac{\partial M_{DB}}{\partial R_1} dx_2 = 0$$

or $\quad -\int_0^{L/2} R_1 x_1 (-x_1) \, dx_1$

$$+ \int_0^{L/2} \left[ -R_1 (x_2 + \frac{L}{2}) + w x_2 \right] \left[ -(x_2 + \frac{L}{2}) \right] dx_2 = 0$$

As both integrals are between the same limits, the subscripts for x may be dropped and the integral equation becomes

$$\int_0^{L/2} \left[ (2R_1 - W)(x^2 + \frac{Lx}{2}) + \frac{R_1 L^2}{4} \right] dx = 0$$

or $\quad \frac{5L^3}{48}(2R_1 - W) + \frac{R_1 L^3}{8} = 0$

from which

$$R_1 = \frac{5W}{16}$$

Applying the equations of statical equilibrium to the half beam, we have

$$R_1 + \frac{R_2}{2} = W \quad \text{or} \quad R_2 = \frac{11W}{8}$$

and  $M + R_1L = \frac{WL}{2}$  or  $M = \frac{3WL}{16}$

The final reactions and the bending moment diagram for the beam are shown in figure 3.5.

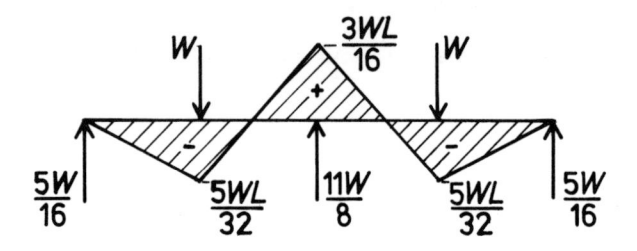

Figure 3.5

## 3.3 CURVED MEMBERS

Beams may be curved and frames may consist of curved members or a combination of straight and curved members. Attention will be confined here to the type of structure in which the plane of loading coincides with the plane of curvature.

The curved structural members most often encountered are the arch, and closed forms such as the circular ring and the chain link. It is essential that these curved members have a radius of curvature which is large in comparison with the depth of the cross-section. If this is not so, the simple theory of bending does not apply, nor can the contributions of axial and shear forces to the total strain energy be neglected.

### 3.3.1 Statically Determinate Curved Members

Rings, arches and links are usually statically indeterminate, but there are a few simpler, statically determinate, shapes which may be examined as an introduction to the analysis of more complex structures consisting of curved members. One such is the davit which is the subject of the next example.

*Example 3.5*

A davit consists of a circular quadrant mounted on a column as shown in figure 3.6. Determine the vertical and horizontal deflexions of the load point A when it is subjected to a vertical load W. The davit has a constant flexural rigidity, EI throughout.

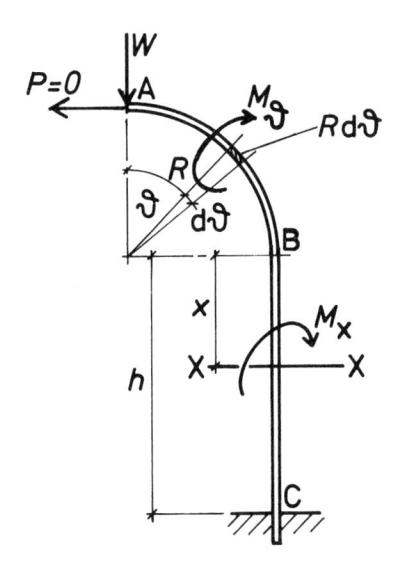

Figure 3.6

Figure 3.6 also shows the dummy horizontal load P which is required to determine the horizontal deflexion at A.

Before we can evaluate the complementary strain energy stored in the davit it is necessary to determine the bending moments in the quadrant AB and the column BC.

For the quadrant, at an element of arc, R dθ, displaced an angle θ from A, the bending moment is given by

$$M_\theta = WR \sin\theta + PR(1 - \cos\theta) \qquad \text{(i)}$$

thus the complementary bending strain energy stored in the quadrant is is given by

$$C_{quad} = \int_0^{\pi/2} \frac{M_\theta^2}{2EI} R \, d\theta \qquad \text{(ii)}$$

For the column, at an element of length, dx, distance x from B, the bending moment is given by

$$M_x = WR + P(R + x) \qquad \text{(iii)}$$

thus the complementary bending strain energy stored in the column is given by

$$C_{col} = \int_0^h \frac{M_x^2}{2EI} \, dx \qquad \text{(iv)}$$

The total complementary energy in the davit is thus

$$C = C_{quad} + C_{col} \tag{v}$$

To determine the deflexions we make us of the first theorem of complementary strain energy, from which

$$\frac{\partial C}{\partial P} = \Delta_H \quad \text{and} \quad \frac{\partial C}{\partial W} = \Delta_V \tag{vi}$$

where $\Delta_H$ and $\Delta_V$ are respectively the horizontal and vertical deflexions at A. Thus from equations (ii), (iv), (v) and (vi)

$$EI\Delta_H = \int_0^{\pi/2} M_\theta \frac{\partial M_\theta}{\partial P} R\, d\theta + \int_0^h M_x \frac{\partial M_x}{\partial P}\, dx \tag{vii}$$

and

$$EI\Delta_V = \int_0^{\pi/2} M_\theta \frac{\partial M_\theta}{\partial W} R\, d\theta + \int_0^h M_x \frac{\partial M_x}{\partial W}\, dx \tag{viii}$$

From equations (i) and (iii) we obtain the partial differential coefficients of the bending moments as

$$\frac{\partial M_\theta}{\partial P} = R(1 - \cos\theta), \quad \frac{\partial M_x}{\partial P} = (R + x)$$

and

$$\frac{\partial M_\theta}{\partial W} = R\sin\theta, \quad \frac{\partial M_x}{\partial W} = R$$

Making the appropriate substitutions in equations (vii) and (viii) and noting that P is zero we obtain

$$EI\Delta_H = WR^3 \int_0^{\pi/2} \sin\theta(1 - \cos\theta)\, d\theta + WR^2 \int_0^h (1 + x/R)\, dx$$

and

$$EI\Delta_V = WR^3 \int_0^{\pi/2} \sin^2\theta\, d\theta + WR^2 \int_0^h dx$$

from which

$$\Delta_H = \frac{WR^3}{2EI}(1 + \frac{h}{R})^2$$

and

$$\Delta_V = \frac{WR^3}{EI}\left(\frac{\pi}{4} + \frac{h}{R}\right)$$

*Example 3.6*

A thin steel strip of flexural rigidity EI is formed into an 'S' shape consisting of two semi-circles of radius R as shown in figure 3.7a. The lefthand end, A is pinned to a rigid base and the righthand end, B moves horizontally between a pair of parallel guides. If the member is used as a spring, determine the stiffness when a horizontal force, H is applied at B.

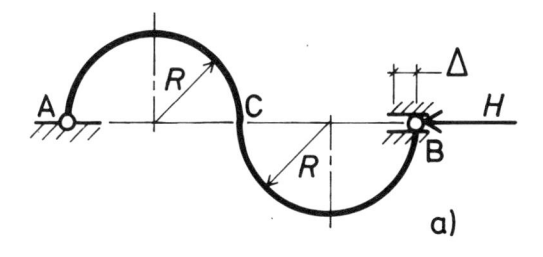

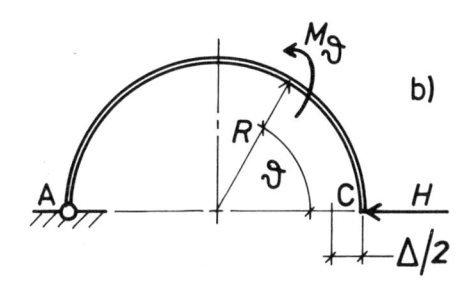

Figure 3.7

Since there is no moment at the pinned support A, there can be no vertical reaction at B and the bending moment at C is therefore also zero. Consequently we need consider only half the member because of symmetry about the mid-point. Figure 3.7b shows the portion of the member from A to C; the force H produces a deflexion $\Delta/2$ at C, where $\Delta$ is the horizontal deflexion at B.

The bending moment in AC, at an element of arc, R d$\theta$ displaced an angle $\theta$ from C is given by

$$M_\theta = HR \sin \theta$$

but from the first theorem of complementary energy

$$\frac{\partial C}{\partial H} = \frac{\Delta}{2} = \int_0^\pi \frac{M_\theta}{EI} \frac{\partial M_\theta}{\partial H} R \, d\theta$$

where

$$\frac{\partial M_\theta}{\partial H} = R \sin \theta$$

thus $\Delta = \dfrac{2HR^3}{EI} \displaystyle\int_0^\pi \sin^2 \theta \, d\theta$

or $\Delta = \dfrac{\pi H R^3}{EI}$

from which the spring stiffness is given by

$$\frac{H}{\Delta} = \frac{EI}{\pi R^3} = 0.32 \, \frac{EI}{R^3}$$

*Example 3.7*

A thin strip of steel of flexural rigidity EI is formed into a circular ring of radius R and loaded as shown in figure 3.8a. If the two ends of the strip are separated by a small distance $\Delta$ when the ring is unloaded, what value of W is required to close the gap?

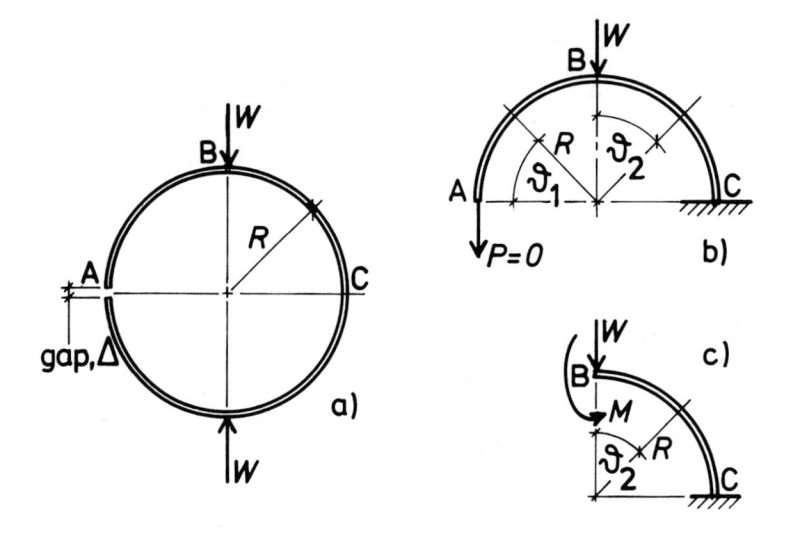

Figure 3.8

Because of the symmetry of the ring and its loading about the horizontal diameter, it is sufficient to consider the half ring AC in figure 3.8b. To determine the vertical deflexion at A, a dummy load, P is applied as shown. The expression for the bending moment is required in two parts since it is a discontinuous function of $\theta$. Thus, from A to B

$$M_{\theta_1} = PR(1 - \cos \theta_1) \qquad 0 < \theta_1 < \frac{\pi}{2}$$

and from B to C

$$M_{\theta_2} = PR(1 + \sin \theta_2) + WR \sin \theta_2 \qquad 0 < \theta_2 < \frac{\pi}{2}$$

To close the gap, the vertical deflexion at A for the half ring should be $\Delta/2$. Thus from the first theorem of complementary energy, we have

$$\frac{\partial C}{\partial P} = \frac{\Delta}{2} = \int_0^{\pi/2} \frac{M_{\theta_1}}{EI} \frac{\partial M_{\theta_1}}{\partial P} R \, d\theta_1 + \int_0^{\pi/2} \frac{M_{\theta_2}}{EI} \frac{\partial M_{\theta_2}}{\partial P} R \, d\theta_2 \qquad (i)$$

where

$$\frac{\partial M_{\theta_1}}{\partial P} = R(1 - \cos \theta_1)$$

and $\quad \dfrac{\partial M_{\theta_2}}{\partial P} = R(1 + \sin \theta_2)$

but since P is zero, the first integral in equation (i) disappears and we have

$$\Delta = \frac{2WR^3}{EI} \int_0^{\pi/2} (\sin \theta_2 + \sin^2 \theta_2) \, d\theta_2$$

or $\quad \Delta = \dfrac{2WR^3}{EI}(1 + \dfrac{\pi}{4})$

hence

$$W = \frac{2EI\Delta}{(\pi + 4)R^3}$$

An alternative approach to this problem is to note that since there is no complementary strain energy stored in the quadrant AB, only the quadrant BC need be considered. The vertical deflexion at A can now be expressed as

$$\frac{\Delta}{2} = \delta + R\phi \qquad (ii)$$

where $\delta$ and $\phi$ are respectively the vertical deflexion and the slope at B.

For the quadrant BC we have

$$\frac{\partial C}{\partial W} = \delta \quad \text{and} \quad \frac{\partial C}{\partial M} = \phi$$

where M is a dummy couple acting at B as shown in figure 3.8c.

The bending moment in BC is therefore given by

$$M_{\theta_2} = M + WR \sin \theta_2$$

hence

$$\frac{\partial C}{\partial W} = \delta = \int_0^{\pi/2} \frac{M_{\theta_2}}{EI} \frac{\partial M_{\theta_2}}{\partial W} R \, d\theta_2$$

and $\quad \dfrac{\partial C}{\partial M} = \phi = \displaystyle\int_0^{\pi/2} \dfrac{M_{\theta_2}}{EI} \dfrac{\partial M_{\theta_2}}{\partial M} R \, d\theta_2$

where

$$\frac{\partial M_{\theta_2}}{\partial W} = R \sin \theta_2 \quad \text{and} \quad \frac{\partial M_{\theta_2}}{\partial M} = 1$$

thus $\quad \delta = \dfrac{WR^3}{EI} \displaystyle\int_0^{\pi/2} \sin^2 \theta_2 \, d\theta_2 = \dfrac{\pi WR^3}{4EI}$

and $\quad \phi = \dfrac{WR^2}{EI} \displaystyle\int_0^{\pi/2} \sin \theta_2 \, d\theta_2 = \dfrac{WR^2}{EI}$

Substituting for $\delta$ and $\phi$ in equation (ii) we have, as before

$$\Delta = 2 \frac{WR^3}{EI}\left(\frac{\pi}{4} + 1\right)$$

or $\quad W = \dfrac{2EI}{(\pi + 4)R^3}$

### 3.3.2 *Statically Indeterminate Curved Members*

The principles involved in dealing with statically indeterminate curved members are no different from those already encountered in the investigation of redundant pin-jointed plane frames and beams. The unknown forces are obtained from equations, derived from the second theorem of complementary energy, which express the compatibility conditions in the structure. Deflexions and rotations, when required, may be obtained by application of the first theorem of complementary energy.

The statically indeterminate arch is often used for spanning large spaces and may be of circular form or, more often, of parabolic form. This type of arch is a continuous structure which is either pinned, or built-in at the abutments. One example of the force analysis of each arch form is given below.

*Example 3.8*

The uniform, two-pinned arch of flexural rigidity EI shown in figure 3.9a has a constant radius of curvature (R) when carrying no load. The abutments A and B are pinned to a rigid foundation at the same level and the angle AOB is 120°, where O is the centre of curvature for the arch.

The arch carries a concentrated load (W) at mid-span and a further load (W) which may be assumed to be uniformly distributed horizontally across the whole span. Taking account of energy due to bending only, determine the horizontal thrust at the supports and sketch the bending moment diagram showing principal values. [Sussex]

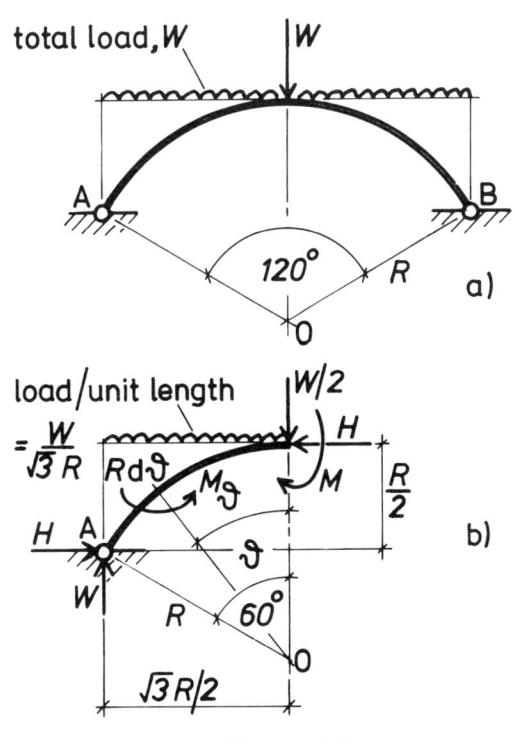

Figure 3.9

Because of the symmetry of the arch, it is sufficient to consider the half span AC shown, together with the forces acting on it, in figure 3.9b.

For equilibrium of the half span, we have

$$M + \frac{W}{2}\frac{\sqrt{3}R}{2} + \frac{W}{2}\frac{\sqrt{3}R}{4} = \frac{HR}{2}$$

or  $8M + 3\sqrt{3}WR = 4HR$ \hfill (i)

If we now consider an element of arc $R\,d\theta$, at an angular distance $\theta$ from C, the bending moment at this point is given by

$$M_\theta = M + \frac{W}{2}R\sin\theta + \frac{W}{\sqrt{3}R}\frac{R\sin\theta}{2}^2 - HR(1 - \cos\theta)$$

or  $$M_\theta = M + \frac{WR}{2}\sin\theta\left(1 + \frac{\sin\theta}{\sqrt{3}}\right) - HR(1 - \cos\theta) \hfill (ii)$$

Either M or H may be considered as the redundancy for this system. If we choose M, the working is slightly simpler. From the second theorem of complementary energy we have

$$\frac{\partial C}{\partial M} = 2\int_0^{\pi/3} \frac{M_\theta}{EI}\frac{\partial M_\theta}{\partial M} R\,d\theta = 0$$

where, from equation (ii), $\partial M_\theta/\partial M = 1$. Hence

$$\int_0^{\pi/3}\left[M + \frac{WR}{2}\sin\theta\left(1 + \frac{\sin\theta}{\sqrt{3}}\right) - HR(1 - \cos\theta)\right]d\theta = 0$$

thus $\left[(M - HR)\theta - \frac{WR}{2}\cos\theta + \frac{WR}{2\sqrt{3}}\left(\frac{\theta}{2} - \frac{\sin 2\theta}{4}\right) + HR\sin\theta\right]_0^{\pi/3} = 0$

or  $16\pi\sqrt{3}M + 8\sqrt{3}HR[3\sqrt{3} - 2\pi] + WR[9\sqrt{3} + 4\pi] = 0$ \hfill (iii)

From equations (i) and (iii) we may solve for M and H, giving

$$H = \frac{W}{8\sqrt{3}}\frac{(14\pi - 9\sqrt{3})}{[3\sqrt{3} - \pi]} = 0.997W \simeq W$$

and $M = \frac{WR}{16}\left(\frac{63\sqrt{3} - 32\pi}{\pi\sqrt{3} - 9}\right) = -0.151WR$

The bending moments in the arch are obtained by substituting for H and M in equation (ii), then

$$M_\theta = WR\left[\frac{\sin\theta}{2}\left(1 + \frac{\sin\theta}{\sqrt{3}}\right) + \cos\theta - 1.15\right]$$

94

from which the bending moment diagram shown in figure 3.10 is obtained.

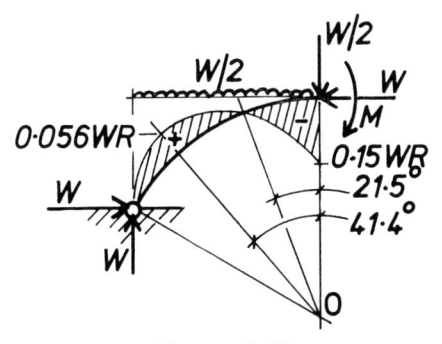

Figure 3.10

The maximum positive value of the bending moment occurs at a value of $\theta$ given by

$$\frac{dM_\theta}{d\theta} = WR\left(\frac{\cos\theta}{2} + \frac{\sin 2\theta}{2\sqrt{3}} - \sin\theta\right) = 0$$

or $\cos\theta + \dfrac{\sin 2\theta}{\sqrt{3}} = 2\sin\theta$

By making the substitutions

$$\cos\theta = \frac{1-t^2}{1+t^2} \quad \text{and} \quad \sin\theta = \frac{2t}{1+t^2}$$

where $t = \tan\theta/2$, we obtain a quartic in t which may be solved by trial to give the desired root in the range $0<\theta<\pi/3$ as

$t = 0.378$

hence

$\theta = 0.723$ rad or $41.4°$

The bending moment in the arch is zero when

$$\frac{\sin\theta}{2}\left(1 + \frac{\sin\theta}{\sqrt{3}}\right) + \cos\theta = 1.15$$

Substituting for $\sin\theta$ and $\cos\theta$ as above we obtain another quartic in t which is solved by trial. There are two roots in the range, $0<\theta<\pi/3$

$t_1 = 0.577$ and $t_2 = 0.1896$

from which

$\theta_1 = 1.05$ rad or $60°$ as expected

95

and

$\theta_2 = 0.375$ rad or $21.5°$

The analysis of parabolic arches of uniform cross-section requires the evaluation of quite complex integrals because of the need to integrate along the centre-line of the arch. Since this evaluation can be time-consuming, examination questions on this type of arch usually deal with a non-uniform cross-section which permits integration to be carried out along the chord of the arch. The following example is typical of this sort of problem.

*Example 3.9*

The symmetrical parabolic arch of span L and height L/4 has a second moment of area given by

$I = I_0 \sec \alpha$

where $\alpha$ is the slope of the arch centre line and $I_0$ is the second moment of area at the crown.

One abutment is pinned and the other is built-in. If the loading consists of a single concentrated load W at the crown, determine the bending moment diagram for the complete arch.

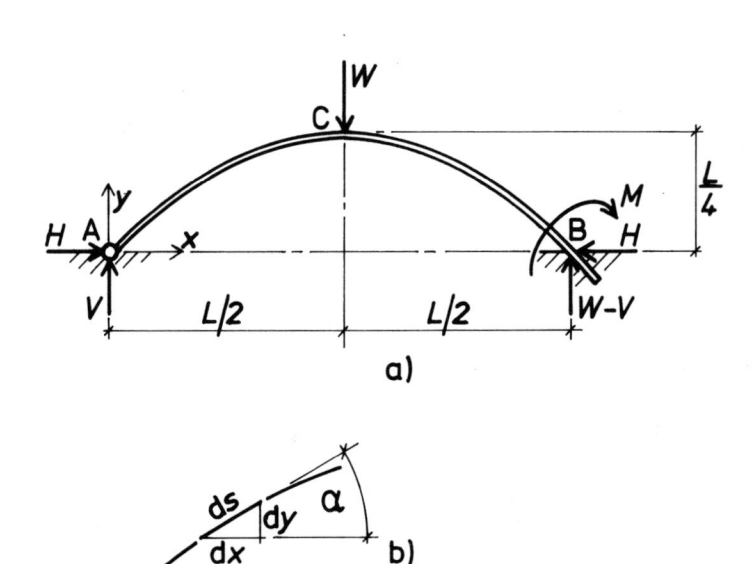

Figure 3.11

Figure 3.11a shows the loaded arch and the reactions at the abutments. Although the arch itself is symmetrical about the crown, the boundary conditions are not and we must consider the complete arch when applying the second theorem of complementary energy.

The equation of the arch centre-line is given by

$$y = \frac{x}{L}(L - x) \tag{i}$$

where y is the height at a horizontal distance x from the left-hand abutment A.

For an element ds of the arch centre-line at x and y from A, we have

$$M_{AC} = Hy - Vx \qquad 0<x<\frac{L}{2}$$

and $\quad M_{CB} = Hy - Vx + W(x - \frac{L}{2}) \qquad \frac{L}{2}<x<L$

There are two redundant forces, H and V. From the second theorem of complementary energy we have

$$\frac{\partial C}{\partial H} = \int_0^{S/2} \frac{M_{AC}}{EI} \frac{\partial M_{AC}}{\partial H} ds + \int_{S/2}^{S} \frac{M_{BC}}{EI} \frac{\partial M_{BC}}{\partial H} ds = 0 \tag{ii}$$

and $\quad \dfrac{\partial C}{\partial V} = \displaystyle\int_0^{S/2} \dfrac{M_{AC}}{EI} \dfrac{\partial M_{AC}}{\partial V} ds + \int_{S/2}^{S} \dfrac{M_{BC}}{EI} \dfrac{\partial M_{BC}}{\partial V} ds = 0 \tag{iii}$

where S is the length of the arch centre-line.

Since the moments above are functions of x, it is necessary to express the element ds in terms of an equivalent horizontal element dx. We note from figure 3.11b that a relationship between ds and dx is given by

$$ds = \sec \alpha \, dx$$

But the cross-section of the arch is such that

$$I = I_0 \sec \alpha$$

thus $ds = \dfrac{I}{I_0} dx$

where $I_0$ is a constant.

Substituting for ds in equations (ii) and (iii) above and changing the limits of integration, we obtain

97

$$\int_0^{L/2} M_{AC} \frac{\partial M_{AC}}{\partial H} dx + \int_{L/2}^{L} M_{BC} \frac{\partial M_{BC}}{\partial H} dx = 0 \qquad (iv)$$

and $$\int_0^{L/2} M_{AC} \frac{\partial M_{AC}}{\partial V} dx + \int_{L/2}^{L} M_{BC} \frac{\partial M_{BC}}{\partial V} dx = 0 \qquad (v)$$

From the bending moment expressions

$$\frac{\partial M_{AC}}{\partial H} = \frac{\partial M_{BC}}{\partial H} = y.$$

and $$\frac{\partial M_{AC}}{\partial V} = \frac{\partial M_{BC}}{\partial V} = -x$$

thus from equations (iv) and (v) we obtain

$$\int_0^{L/2} (Hy - Vx)y\, dx + \int_{L/2}^{L} [Hy - Vx + W(x - \tfrac{L}{2})]y\, dx = 0 \qquad (vi)$$

$$\int_0^{L/2} (Hy - Vx)(-x)\, dx + \int_{L/2}^{L} [Hy - Vx + W(x - \tfrac{L}{2})](-x)\, dx = 0 \qquad (vii)$$

After making the substitution for y from equation (i) and integrating, equations (vi) and (vii) become

$$\frac{H}{15} - \frac{V}{6} + \frac{W}{32} = 0$$

and $$H - 4V + \frac{5W}{4} = 0$$

From which

$$H = \frac{5W}{6}$$

and $$V = \frac{25W}{48}$$

The bending moments in AC and BC are now given by

$$M_{AC} = \frac{5WL}{48}\left(\frac{x}{L}\right)\left[3 - 8\left(\frac{x}{L}\right)\right]$$

and $$M_{BC} = -\frac{WL}{48}\left[24 - 63\left(\frac{x}{L}\right) + 40\left(\frac{x}{L}\right)^2\right]$$

These expressions are used to draw the non-dimensionalised bending moment diagram shown in figure 3.12.

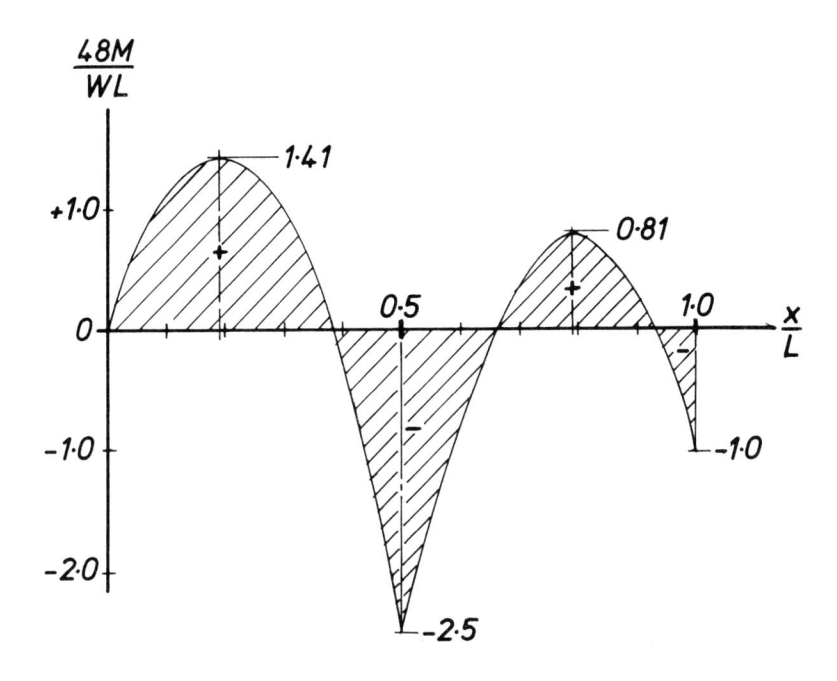

Figure 3.12

The maximum positive bending moments are 1.41 (WL/48) and 0.81 (WL/48) and occur at 0.19L and 0.79L from A.

The largest negative moment is 2.5 (WL/48) and occurs at the crown.

The bending moment is zero at x = 0, 0.375L, 0.645L and 0.93L.

Closed rings and links, although not often used as structural members, are included here for completeness. A classic problem of this type concerns the circular ring loaded across a diameter.

*Example 3.10*

Determine the distribution of bending moment in the thin ring of uniform section and radius R shown in figure 3.13a. The ring is loaded by a pair of equal and opposite tensile forces acting across a diameter. Determine also the increase in the ring diameter in the line of action of the load and the decrease in the diameter at right-angles to this direction.

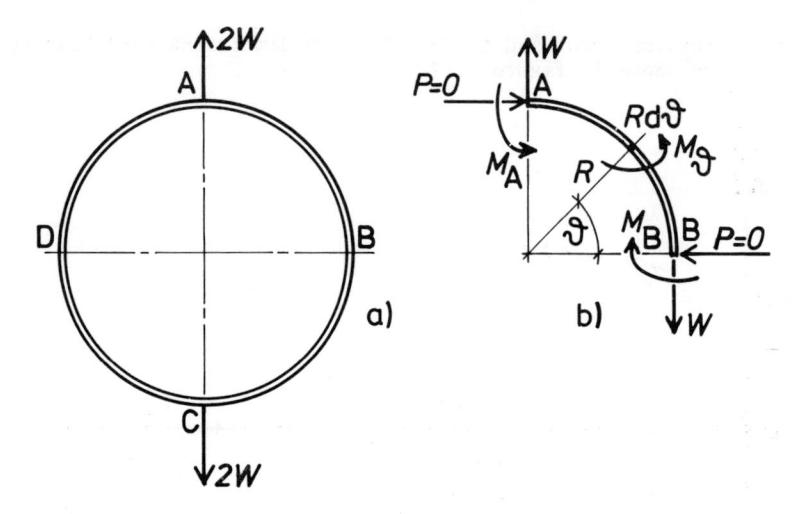

Figure 3.13

Since the ring has two-fold symmetry, it is sufficient to consider the quadrant AB shown in figure 3.13b. The external forces and the internal reactions are also shown in the figure, together with the dummy load P at B needed to determine the decrease in the horizontal diameter. Again because of symmetry, the axial force at A and the shear force at B are both zero (although P must be added as shown). The only redundancy in the quadrant is either the moment at A or the moment at B.

Choosing $M_B$ as the redundancy we find that the moment at an angle $\theta$ from B is given by

$$M_\theta = M_B + WR(1 - \cos \theta) + PR \sin \theta \qquad (i)$$

From the second theorem of complementary energy we have

$$\frac{\partial C}{\partial M_B} = 4 \int_0^{\pi/2} \frac{M_\theta}{EI} \frac{\partial M_\theta}{\partial M_B} R \, d\theta = 0$$

then, since P is zero

$$\int_0^{\pi/2} \left[ M_B + WR(1 - \cos \theta) \right] \times 1 \times R \, d\theta = 0$$

or 
$$\left[ (M_B + WR)\theta - WR \sin \theta \right]_0^{\pi/2} = 0$$

hence

$$M_B = -WR\left(1 - \frac{2}{\pi}\right)$$

The negative sign indicates that $M_B$ is a moment tending to decrease the curvature at B.

The distribution of bending moment in the ring may now be obtained by substituting for $M_B$ in equation (i) noting that P is zero, thus

$$M_\theta = WR\left(\frac{2}{\pi} - \cos\theta\right) \qquad \text{(ii)}$$

Figure 3.14 shows the moments in the quadrant determined from equation (ii).

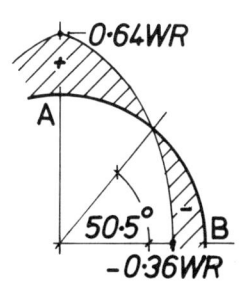

Figure 3.14

The increase, $\Delta_{AC}$ in the diameter, AB is given by

$$\Delta_{AC} = 2\delta_A$$

where $\delta_A$ is the vertical deflexion at A in the quadrant AB.

Similarly the decrease, $\Delta_{BD}$ in the horizontal diameter BD is

$$\Delta_{BD} = -2\delta_B$$

where $\delta_B$ is the horizontal deflexion at B in the quadrant AB.

From the first theorem of complementary energy we have, for the quadrant

$$\frac{\partial C}{\partial W} = \int_0^{\pi/2} \frac{M_\theta}{EI} \frac{\partial M_\theta}{\partial W} R\, d\theta = \delta_A \qquad \text{(iii)}$$

and $\quad \dfrac{\partial C}{\partial P} = \displaystyle\int_0^{\pi/2} \dfrac{M_\theta}{EI} \dfrac{\partial M_\theta}{\partial P} R\, d\theta = \delta_B$ \hfill (iv)

where

$$\dfrac{\partial M_\theta}{\partial W} = R\left(\dfrac{2}{\pi} - \cos\theta\right)$$

and $\quad \dfrac{\partial M_\theta}{\partial P} = R\sin\theta$

therefore, since P is zero, equations (iii) and (iv) become

$$\delta_A = \dfrac{WR^3}{EI} \int_0^{\pi/2} \left(\dfrac{2}{\pi} - \cos\theta\right)^2 d\theta$$

and $\quad \delta_B = \dfrac{WR^3}{EI} \displaystyle\int_0^{\pi/2} \left(\dfrac{2}{\pi} - \cos\theta\right) \sin\theta\, d\theta$

hence

$$\delta_A = \dfrac{WR^3}{4\pi EI}(\pi^2 - 8)$$

and $\quad \delta_B = \dfrac{WR^3}{2\pi EI}(4 - \pi)$

Thus the changes in the diameters are

$$\Delta_{AC} = \dfrac{WR^3}{2\pi EI}(\pi^2 - 8) = +0.30\,\dfrac{WR^3}{EI}$$

and $\quad \Delta_{BD} = -\dfrac{WR^3}{\pi EI}(4-\pi) = -0.27\,\dfrac{WR^3}{EI}$

The next example is a little more difficult because of the need to determine the bending moment due to pressure acting on a curved surface.

*Example 3.11*

A tube having a uniform thickness of 100 mm has a cross-section as shown in figure 3.15a. If the maximum tensile stress in the tube wall is not to exceed 120 MN m$^{-2}$ determine the maximum internal pressure which may be applied to the tube. Ignore end effects and take account of bending energy only.

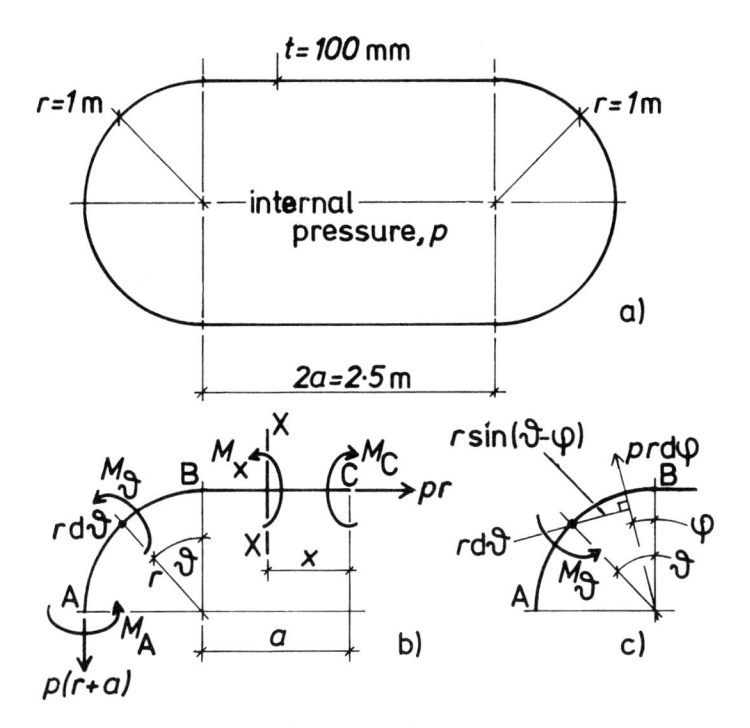

Figure 3.15

The tube has two-fold symmetry, thus it is sufficient to consider a quarter of the tube represented by the davit shape ABC shown in figure 3.15b. If unit length of tube is considered, the pressure p may be treated as a uniformly distributed load. Notice also that due to symmetry, there are no shear forces at A or C.

In the portion of the tube wall from C to B the bending moment is given by

$$M_x = M_C - \frac{px^2}{2} \tag{i}$$

and from B to A the moment is

$$M_\theta = M_C + pr^2(1 - \cos\theta) - pa(\frac{a}{2} + r\sin\theta) - \int_0^\theta pr^2 \sin(\theta - \phi)\, d\phi$$

The last term in this expression represents the bending moment due to the pressure on the curved part of the tube. It may be derived by reference to figure 3.15c.

After evaluating the integral, we find that the second and last terms in the expression for $M_\theta$ cancel, thus

$$M_\theta = M_C - pa(\frac{a}{2} + r\sin\theta) \tag{ii}$$

The second theorem of complementary energy may now be applied to determine the unknown moment $M_C$ in equations (i) and (ii). Thus

$$\frac{\partial C}{\partial M_C} = \frac{4}{EI}\int_0^a M_x \frac{\partial M_x}{\partial M_C} dx + \frac{4}{EI}\int_0^{\pi/2} M_\theta \frac{\partial M_\theta}{\partial M_C} r\, d\theta = 0$$

where

$$\frac{\partial M_x}{\partial M_C} = \frac{\partial M_\theta}{\partial M_C} = 1$$

hence

$$\int_0^a (M_C - \frac{px^2}{2})\, dx + \int_0^{\pi/2}\left[M_C - pa(\frac{a}{2} + r\sin\theta)\right] r\, d\theta = 0$$

or 
$$M_C = \frac{pa}{12}\left(\frac{4a^2 + 6\pi ar + 24r^2}{2a + \pi r}\right)$$

Evaluating the bending moments in terms of the pressure p for the values of r and a given, we have

$$M_C = 0.994p$$

$$M_x = (0.994 - \frac{x^2}{2})p$$

and $M_\theta = (0.212 - 1.25\sin\theta)p$

These expressions may be used to draw the bending moment diagram for the quarter tube shown in figure 3.16. The units for the moments are MN m if the pressure is in MN m$^{-2}$.

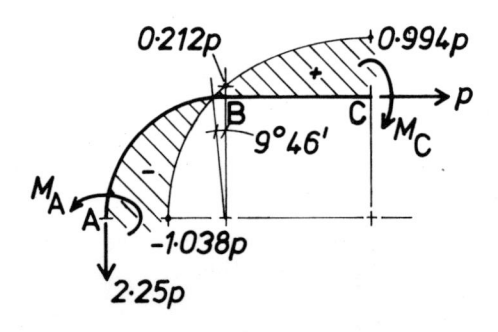

Figure 3.16

Also shown in figure 3.16 are the tensile forces per unit length of tube acting at A and C. These forces are in MN if p is in MN m$^{-2}$.

The greatest tensile stress in the tube wall will clearly occur on the inner surface at A, hence

$$\sigma_{max} = 1.038p \frac{6}{t^2} + \frac{2.25p}{t} \text{ MN m}^{-2} \qquad \text{(iii)}$$

where t is the wall thickness in m; but $\sigma_{max}$ is not to exceed 120 MN m$^{-2}$, thus

$$p_{max} = \frac{120}{645.3} \text{ MN m}^{-2} = 1.86 \text{ bar}$$

If, for example, the tube had been required to withstand a maximum internal pressure of 2 bar, equation (iii) could be used to determine that the minimum wall thickness required is 104 mm.

## 3.4 RIGID-JOINTED PLANE FRAMES

If the analysis of forces in rigid-jointed plane frames is required, application of the second theorem of complementary energy will provide as many equations as there are redundant forces in the structure. If the structure is a complex multi-storey, multi-bay frame the number of equations for the redundants are likely to be too great for solution by hand and the resources of a computer may have to be called upon. However, there is no fundamental difference in the procedure for setting up these equations whatever the number of redundancies. Thus for the purposes of illustration, it will be sufficient to examine frames with not more than three redundancies. Once forces have been obtained, the first theorem of complementary energy may be used to determine each deflexion associated with the structure.

It must be recognised that there are a number of alternative methods for the analysis of rigid-jointed frames such as moment distribution or the use of slope-deflexion equations, but these methods are not our concern here. It is possible, however, that these alternatives might be more efficient for the solution of a particular problem. The reader is advised to become conversant with these other methods and to make a careful choice of the procedures available before embarking on a solution.

Since there are no new principles involved, we may proceed directly to illustrating the energy method of analysis for rigid-jointed frames by means of a number of examples.

*Example 3.12*

A closed frame ABCDE of uniform cross-section is shown in figure 3.17a. The frame is constructed from a straight length of bar which is bent into shape and the ends pin-connected at A. The support at D can resist horizontal and vertical forces only. The support at C, being a roller bearing, can only provide a vertical reaction.

Determine the bending moment at the centre of the span DC when a horizontal load W is applied at F as shown.

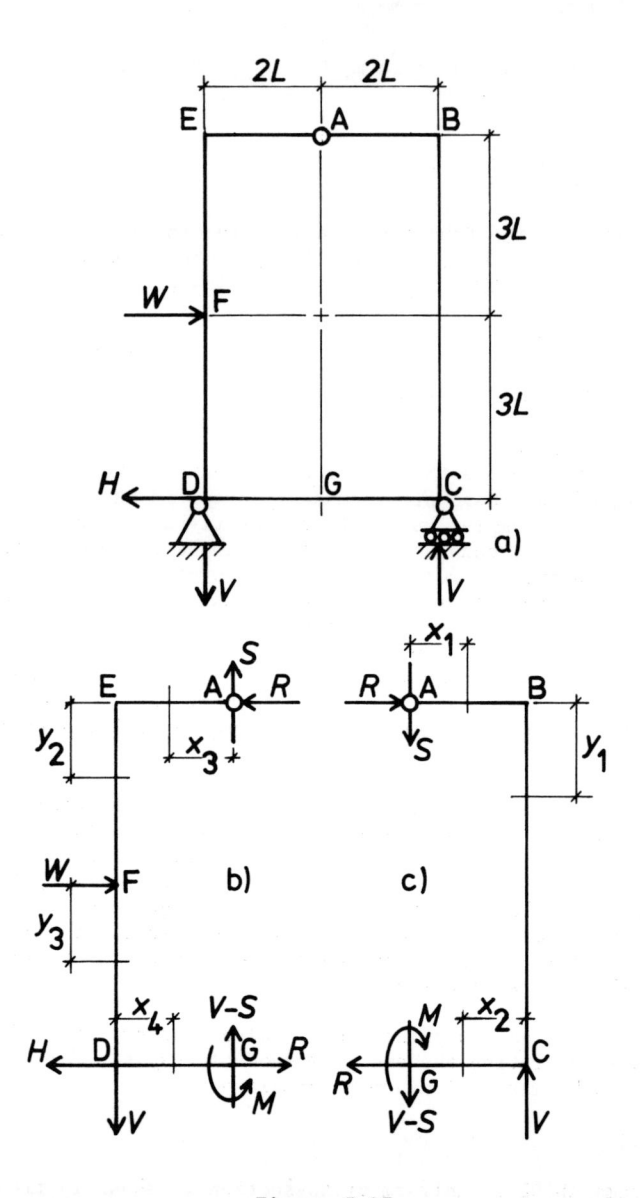

Figure 3.17

The frame is statically determinate with respect to the reactions, thus

$$H = W \quad \text{and} \quad V = \frac{3W}{4}$$

There is no axis of symmetry for the loading, thus the whole frame must be considered. We note that since there is a pin joint at A there can only be shear and axial forces acting at this point. If the frame is separated at the vertical centre-line, the equilibrium of each half may be maintained by applying a shear force and an axial force at A and a shear force, an axial force and a moment at G. The moment at G may be expressed in terms of the axial force at A, thus the redundant forces in the system are the forces S and R at A. Figures 3.17b and c show the forces acting on each half of the frame.

The moment M at the centre of the span DC may be obtained from either of figures 3.17b or c and is given by

$$M = \frac{3L}{2}(W - 4R)$$

The bending moments in each section of the frame may be obtained from figures 3.17b and c as follows.

$$M_{AB} = x_1 S \qquad 0 < x_1 < 2L$$

$$M_{BC} = 2LS - y_1 R \qquad 0 < y_1 < 6L$$

$$M_{CG} = (2L - x_2)S - 6RL + V x_2 \qquad 0 < x_2 < 2L$$

$$M_{AE} = -x_3 S \qquad 0 < x_3 < 2L$$

$$M_{EF} = -2LS - R y_2 \qquad 0 < y_2 < 3L$$

$$M_{FD} = -2LS - R(3L + y_3) + W y_3 \qquad 0 < y_3 < 3L$$

$$M_{DG} = -(2L - x_4)S - 6RL + 3WL - V x_4 \qquad 0 < x_4 < 2L$$

Appling the second theorem of complementary energy to the whole frame, we have

$$\frac{\partial C}{\partial R} = \frac{\partial C}{\partial S} = 0$$

thus $\displaystyle\int_0^{2L} \frac{M_{AB}}{EI} \frac{\partial M_{AB}}{\partial R} dx_1 + \int_0^{6L} \frac{M_{BC}}{EI} \frac{\partial M_{BC}}{\partial R} dy_1$

$$+ \int_0^{2L} \frac{M_{CG}}{EI} \frac{\partial M_{CG}}{\partial R} dx_2 + \int_0^{2L} \frac{M_{AE}}{EI} \frac{\partial M_{AE}}{\partial R} dx_3$$

$$+ \int_0^{3L} \frac{M_{EF}}{EI} \frac{\partial M_{EF}}{\partial R} dy_2 + \int_0^{3L} \frac{M_{FD}}{EI} \frac{\partial M_{FD}}{\partial R} dy_3$$

$$+ \int_0^{2L} \frac{M_{DG}}{EI} \frac{\partial M_{DG}}{\partial R} dx_4 = 0 \qquad (i)$$

and
$$\int_0^{2L} \frac{M_{AB}}{EI} \frac{\partial M_{AB}}{\partial S} dx_1 + \int_0^{6L} \frac{M_{BC}}{EI} \frac{\partial M_{BC}}{\partial S} dy_1$$

$$+ \int_0^{2L} \frac{M_{CG}}{EI} \frac{\partial M_{CG}}{\partial S} dx_2 + \int_0^{2L} \frac{M_{AE}}{EI} \frac{\partial M_{AE}}{\partial S} dx_3$$

$$+ \int_0^{3L} \frac{M_{EF}}{EI} \frac{\partial M_{EF}}{\partial S} dy_2 + \int_0^{3L} \frac{M_{FD}}{EI} \frac{\partial M_{FD}}{\partial S} dy_3$$

$$+ \int_0^{2L} \frac{M_{DG}}{EI} \frac{\partial M_{DG}}{\partial S} dx_4 = 0 \qquad (ii)$$

where

$$\frac{\partial M_{AB}}{\partial R} = 0 \qquad \frac{\partial M_{AB}}{\partial S} = x_1$$

$$\frac{\partial M_{BC}}{\partial R} = y_1 \qquad \frac{\partial M_{BC}}{\partial S} = 2L$$

$$\frac{\partial M_{CG}}{\partial R} = -6L \qquad \frac{\partial M_{CG}}{\partial S} = (2L - x_2)$$

$$\frac{\partial M_{AE}}{\partial R} = 0 \qquad \frac{\partial M_{AE}}{\partial S} = -x_3$$

$$\frac{\partial M_{EF}}{\partial R} = -y_2 \qquad \frac{\partial M_{EF}}{\partial S} = -2L$$

$$\frac{\partial M_{FD}}{\partial R} = -(3L + y_3), \qquad \frac{\partial M_{FD}}{\partial S} = -2L$$

$$\frac{\partial M_{DG}}{\partial R} = -6L \qquad \frac{\partial M_{DG}}{\partial S} = -(2L - x_4)$$

After substituting these values in equation (i) and integrating, we find that S cancels and

$$R = \frac{13W}{64}$$

Since the desired moment M is independent of the shear force S, there is no need to make use of equation (ii) and we have

$$M = \frac{9WL}{32}$$

*Example 3.13*

A rigidly jointed plane frame ABCD, loaded as shown in figure 3.18, is pinned to the foundations and is constructed from members having a flexural rigidity, EI of 500 kN m$^{-2}$.

The horizontal force H applied at joint B prevents the frame from swaying. Determine the magnitude of H and sketch the bending moment diagram for the frame showing the principle moment values and also the horizontal and vertical reactions at the feet.

What would have been the horizontal deflexion of joint B if the force H had not been present? [Sussex]

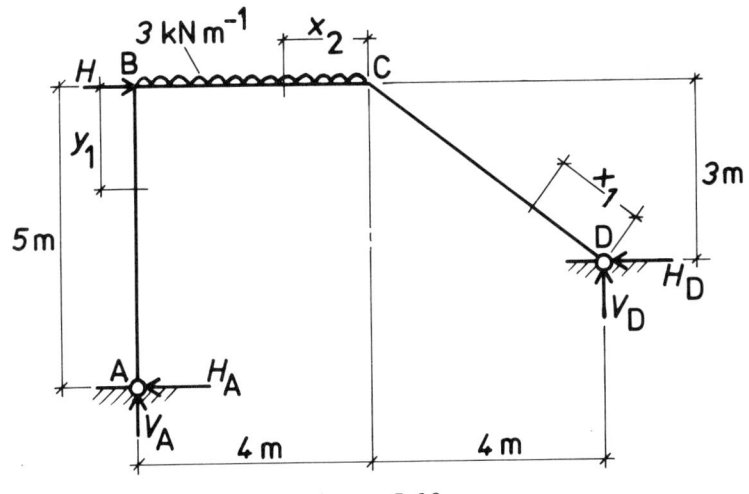

Figure 3.18

From the equations of statical equilibrium

$$V_A + V_D = 12 \text{ kN} \tag{i}$$

$$H_A + H_D = H \tag{ii}$$

and  $2H_D - 5H + 8V_D = 24 \text{ kN m}$

or  $$V_D = 3 + \frac{5H - 2H_D}{8} \tag{iii}$$

There are a total of five unknown forces, thus we require two additional equations to be provided by the compatibility conditions at B and D. From the first and second theorems of complementary energy we have

$$\frac{\partial C}{\partial H} = \Delta_B \quad \text{and} \quad \frac{\partial C}{\partial H_D} = 0 \tag{iv}$$

where $H$ and $H_D$ are chosen as the redundant forces.

The bending moments in the frame are

$$M_{DC} = (3H_D - 4V_D)\frac{x_1}{5} \quad 0<x_1<5 \text{ m}$$

$$M_{CD} = 3H_D - (4 + x_2)V_D + \frac{3}{2}x_2^2 \quad 0<x_2<4 \text{ m}$$

and  $M_{BA} = (3 - y_1)H_D - 8V_D + 24 + Hy_1 \quad 0<y_1<5 \text{ m}$

From equation (iv) we have

$$\Delta_B = \int_0^5 \frac{M_{DC}}{EI} \frac{\partial M_{DC}}{\partial H} dx_1 + \int_0^4 \frac{M_{CB}}{EI} \frac{\partial M_{DC}}{\partial H} dx_2 + \int_0^5 \frac{M_{BA}}{EI} \frac{\partial M_{BA}}{\partial H} dy_1 \tag{v}$$

and  $$\int_0^5 M_{DC} \frac{\partial M_{DC}}{\partial H_D} dx_1 + \int_0^4 M_{CB} \frac{\partial M_{CB}}{\partial H_D} dx_2 + \int_0^5 M_{BA} \frac{\partial M_{BA}}{\partial H_D} dy_1 = 0 \tag{vi}$$

From the moment expressions and equation (iii) we obtain

$$\frac{\partial M_{DC}}{\partial H} = -\frac{4x_1}{5}\frac{\partial V_D}{\partial H} = -\frac{x_1}{2}$$

$$\frac{\partial M_{CB}}{\partial H} = -(4 + x_2)\frac{\partial V_D}{\partial H} = -\frac{5}{8}(4 + x_2)$$

$$\frac{\partial M_{BA}}{\partial H} = -8 \frac{\partial V_D}{\partial H} + y_1 = (y_1 - 5)$$

and $$\frac{\partial M_{DC}}{\partial H_D} = \frac{3x_1}{5} - \frac{4x_1}{5} \frac{\partial V_D}{\partial H_D} = \frac{4x_1}{5}$$

$$\frac{\partial M_{CB}}{\partial H_D} = 3 - (4 + x_2) \frac{\partial V_D}{\partial H_D} = 4 + \frac{x_2}{4}$$

$$\frac{\partial M_{BA}}{\partial H_D} = (3 - y_1) - 8 \frac{\partial V_D}{\partial H_D} = (5 - y_1)$$

After substituting these values in equations (v) and (vi) and integrating we finally obtain

$$304H_D - 265H - 456 = \frac{12EI}{5} \Delta_B \qquad \text{(vii)}$$

and $\quad 449H_D - 380H - 768 = 0 \qquad \text{(viii)}$

Initially the sway deflexion ($\Delta_B$) is zero, thus from equations (vii) and (viii)

$$304H_D - 265H = 456$$

and $\quad 449H_D - 380H = 768$

hence

$$H = 8.27 \text{ kN} \quad \text{and} \quad H_D = 8.71 \text{ kN}$$

If H is zero, equation (viii) gives $H_D = 1.71$ kN and from equation (vii) we obtain

$$\Delta_B = -\frac{80}{3EI} \text{ m}$$

but $\quad EI = 500$ kN m$^2$, thus

$$\Delta_B = -0.0533 \text{ m} = -53.3 \text{ mm}$$

The last example in this section examines the effect of temperature changes in rigidly jointed frames. This type of calculation is particularly important when the stress analysis is required of pipe lines carrying fluids at extreme temperatures.

*Example 3.14*

The plane pipework system shown in figure 3.19a is of uniform cross-section. Assuming the joints are rigid and the ends are fixed against translation and rotation, calculate the magnitude of the maximum bending moment produced by a temperature rise $\theta$. The material of the pipe has a coefficient of expansion $\alpha$ and an elastic modulus E. [Cambridge]

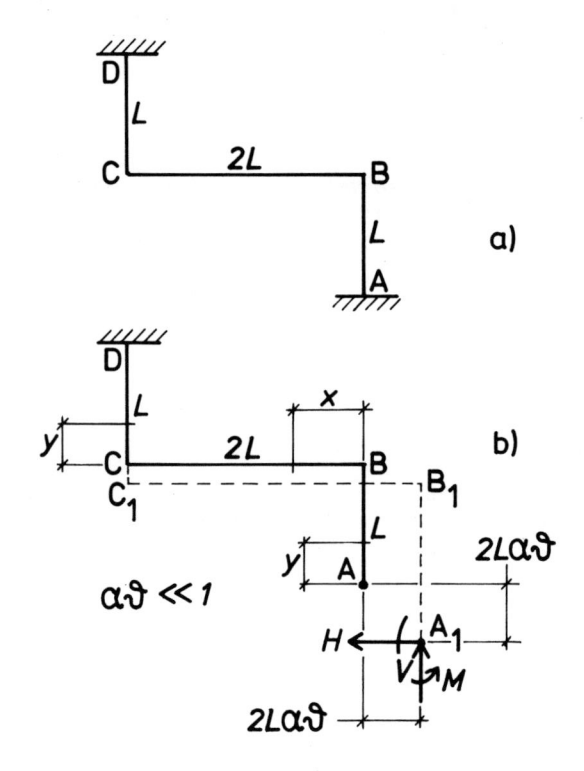

Figure 3.19

If the restraints at A are released, a temperature rise $\theta$ will cause A to move to $A_1$ as shown in figure 3.19b. Also shown in the figure are the forces required to restore $A_1$ to A. The deflexions caused by the horizontal shear H and the thrust V are both equal to $2L\alpha\theta$. The moment M is such that no rotation at A is permitted. Therefore, from the first theorem of complementary energy we have

$$\frac{\partial C}{\partial H} = \frac{\partial C}{\partial V} = 2L\alpha\theta$$

and $\quad \dfrac{\partial C}{\partial M} = 0$

The bending moment expressions for AB, BC and CD are

$$M_{AB} = -M + Hy \qquad 0<y<L$$

$$M_{BC} = -M + HL - Vx \qquad 0<x<2L$$

and $M_{CD} = -M + H(L + y) - 2VL \qquad 0<y<L$

thus $\dfrac{\partial M_{AB}}{\partial H} = y, \quad \dfrac{\partial M_{AB}}{\partial V} = 0, \quad \dfrac{\partial M_{AB}}{\partial M} = -1$

$\dfrac{\partial M_{BC}}{\partial H} = L, \quad \dfrac{\partial M_{BC}}{\partial V} = -x, \quad \dfrac{\partial M_{BC}}{\partial M} = -1$

and $\dfrac{\partial M_{CD}}{\partial H} = (L + y), \quad \dfrac{\partial M_{CD}}{\partial V} = -2L, \quad \dfrac{\partial M_{CD}}{\partial M} = -1$

From the foregoing we obtain the displacement equations

$$\frac{\partial C}{\partial H} = \frac{1}{EI} \int_0^L (Hy^2 - My) \, dy + \frac{L}{EI} \int_0^{2L} (HL - M - Vx) \, dx$$

$$+ \frac{1}{EI} \int_0^L [H(L + y)^2 - M(L + y) - 2VL(L + y)] \, dy = 2L\alpha\theta$$

$$\frac{\partial C}{\partial V} = -\frac{1}{EI} \int_0^{2L} (HLx - Mx - Vx^2) \, dx - \frac{2L}{EI} \int_0^L [H(L + y) - M - 2VL] \, dy$$

$$= 2L\alpha\theta$$

and $\dfrac{\partial C}{\partial M} = -\dfrac{1}{EI} \int_0^L (Hy - M) \, dy - \dfrac{1}{EI} \int_0^{2L} (HL - M - Vx) \, dx$

$$- \frac{1}{EI} \int_0^L [H(L + y) - M - 2VL] \, dy = 0$$

hence

$$14HL - 12M - 15VL = \frac{6EI}{L}\alpha\theta \qquad (i)$$

$$15HL - 12M - 20VL = -\frac{6EI}{L}\alpha\theta \qquad (ii)$$

and  HL - M - VL = 0 (iii)

Equations (i) and (iii) are solved to give

$$H = \frac{66}{7} \frac{EI}{L^2} \alpha\theta$$

$$V = \frac{30}{7} \frac{EI}{L^2} \alpha\theta$$

and  $$M = \frac{36}{7} \frac{EI}{L} \alpha\theta$$

The bending moment diagram for the pipework is shown in figure 3.20 from which the maximum bending moment, which occurs at A and D, is seen to be equal to M.

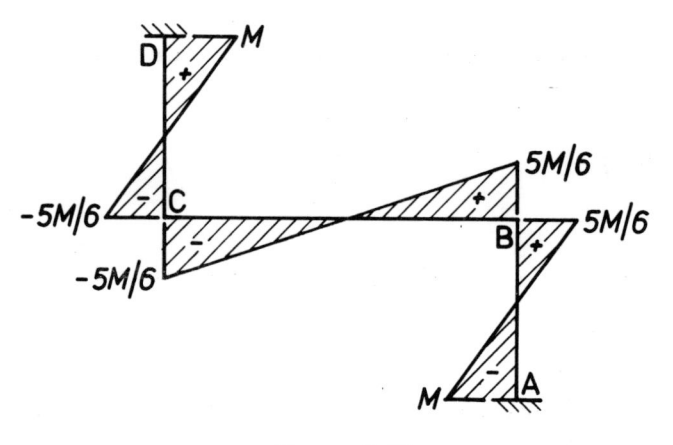

Figure 3.20

## 3.5 DESIGN EXAMPLE

At a certain point in a structure subjected to long-term testing, the maximum applied load in compression is not to exceed 200 kN. The load is applied horizontally using a screw-jack driven by an electric motor through gearing. The motor speed and the gear ratios are chosen so that full load is reached between 80 and 100 hours after the start of a test.

A load cell, placed between the platen of the screw-jack and the structure, provides an output which is converted, electronically, into a continuous record of the applied load. At the same time a linear transducer is used to record the displacement of the load point. The whole testing procedure is thus fully automatic once started.

It is required that the test should stop automatically as soon as the load reaches 200 kN. One way of achieving this is to use the amplified signal from the load cell to switch off the motor driving the screw-jack. However, this method is found to be unreliable because the calibration of the circuitry tends to drift over a long period of time and it is also affected by temperature changes in the laboratory.

It is decided that a mechanical device with a positive action would be more reliable and the design office is requested to offer a solution. After some thought, the designer came up with the basic idea of using a horizontal circular ring placed between the screw-jack and load cell so that a diameter of the ring coincides with the line of action of the load. The change in this diameter can therefore be used as a measure of the load. In itself, this idea is not new, in fact it is the principle behind the proving ring. It is also difficult to see how this deflexion can be used directly to shut off the screw-jack motor. However, the designer is aware that as the ring diameter in the direction of the load decreases, so the diameter at right-angles increases. He realises that if a pre-stressed bar were to be inserted inside the ring and across the transverse diameter it would drop out under its own weight when the compressive applied load reached a certain critical value. If one end of the bar is hinged to the ring, the other end can be used to operate a cut-off switch as it falls. Furthermore, if the ring and bar are made of the same material, the operation of the device is unaffected by changes in the ambient temperature. The designer adopted the following procedure for proportioning the bar and the ring.

There are two extreme loading conditions. The first is when the load transmitted to the structure is zero and the bar is under maximum pre-stress. The second occurs when the load transmitted to the structure reaches the critical value and the force in the bar is zero. Figure 3.21 shows the ring and bar in the first loading condition. The force in the bar is $F_m$, the mean radius of the ring is R and its thickness is t. Diameter AC is in the line of action of the applied load.

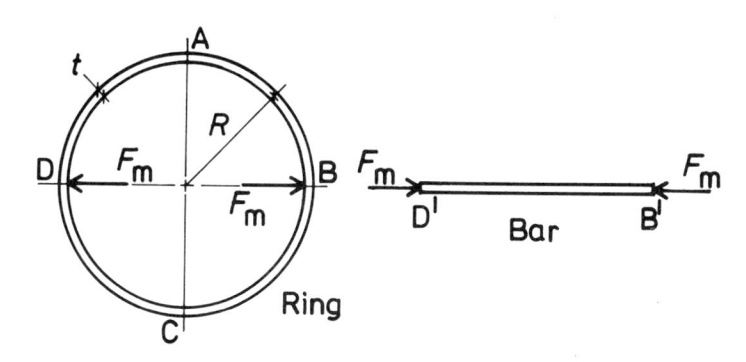

Figure 3.21

Referring to example 3.10, the inside diameter of the ring BD under the action of the forces $F_m$ in the bar is given by

$$BD = (2R - t) + \frac{F_m R^3}{EI}\left(\frac{\pi^2 - 8}{4\pi}\right)$$

where EI is the flexural rigidity of the ring cross-section.

If the initial length of the bar is $(2R - t) + \lambda$, under the action of the axial compressive force $F_m$ the new length B'D' is given by

$$B'D' = (2R - t) + \lambda - \frac{2F_m R}{EA}$$

where A is the cross-sectional area of the bar and 2R is assumed to be large in comparison with $\lambda - t$.

Since the bar is to fit inside the ring we have BD = B'D' or

$$F_m = \frac{EI\lambda}{R^3}\left[\frac{4\pi AR^2}{(\pi^2 - 8)AR^2 + 8\pi I}\right] \tag{i}$$

Figure 3.22 shows the ring and bar in the second loading condition, $W_c$ is the critical, or tripping load for the device.

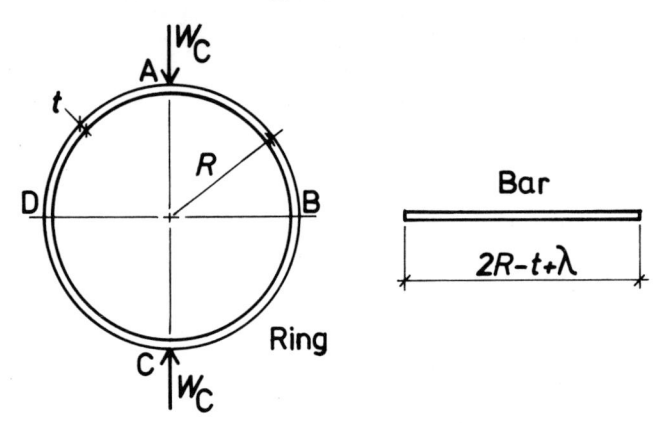

Figure 3.22

There is no load in the bar when the ring diameter BD is equal to its initial length, $(2R - t) + \lambda$. From example 3.10, the diameter BD is given by

$$BD = (2R - t) + \frac{W_c R^3}{EI}\left(\frac{4 - \pi}{2\pi}\right)$$

hence

$$W_c = \frac{EI\lambda}{R^3}\left(\frac{2\pi}{4-\pi}\right) \qquad \text{(ii)}$$

Steel, Grade 55 to British Standard 4360: Part 2: 1969 Weldable Structural Steels, is selected as the material for the ring and bar. E may therefore be taken as $200 \times 10^6$ kN m$^{-2}$. The cross-section of the ring is chosen to be rectangular, of breadth b and thickness, t. The radius, R is taken to be 10 times the thickness.

From equations (i) and (ii) we have

$$F_m = \frac{0.918 W_c}{(1+K)} \qquad \text{(iii)}$$

$$K = 0.0112 bt/A \qquad \text{(iv)}$$

and $\lambda b = 8.2 W_c$ mm$^2$ (v)

if $W_c$ is in kN.

If the bar is made of solid circular section of diameter d, the maximum axial compressive stress is obtained from equation (iii) as

$$\sigma_m = \frac{F_m}{A} = \frac{1.169}{(1+K)} \frac{W_c}{d^2} \qquad \text{(vi)}$$

Since the bar is under axial compression, the possibility of buckling must be considered. In order to determine safe loads for what is effectively a pin-ended strut, it is necessary to calculate the slenderness ratio (L/r) for the bar. The length L is approximately 2R and the radius of gyration r for a circular cross-section of diameter d is 0.25d, thus

$$\frac{L}{r} = \frac{8R}{d} = \frac{80t}{d} \qquad \text{(vii)}$$

From the bending moment diagram of example 3.10 and noting from equation (iii) that $W_c$ is always greater than $F_m$, we find that the greatest bending stress in the ring occurs at A and C under the second loading condition. This stress is given by

$$\sigma_b = \pm 19.2 \frac{W_c}{bt}$$

Also acting at A and C is a shear stress having a maximum value given by

$$\tau_m = 0.75 \frac{W_c}{bt}$$

The equivalent maximum stress $\sigma_e$ at A and C due to combined bending and shear is obtained from Clause 14c of British Standard 449: Part 2: 1969 The Use of Structural Steel in Building, as

$$\sigma_e = \left(\sigma_b^2 + 3\tau_m^2\right)^{\frac{1}{2}} = \pm 19.25 \frac{W_c}{bt} \qquad (viii)$$

From Table 1 of the same British Standard, the allowable equivalent stress for plates, sections and bars of Grade 55 steel having a thickness in excess of 40 mm is 360 MN m$^{-2}$. If $W_c$ in equation (viii) is in kN and b and t are in m, we have

$$\sigma_e = 360 \times 10^3 = 19.25 \frac{W_c}{bt} \text{ kN m}^{-2}$$

thus $bt = 0.0535 W_c \times 10^{-3}$ m$^2$

or $bt = 53.5 W_c$ mm$^2$ \qquad (ix)

Suppose that the ratio of the cross-sectional areas of the ring and the bar is n, then

$$bt = nA = n\frac{\pi d^2}{4} \qquad (x)$$

From equations (ix) and (x) we have

$$\frac{W_c}{d^2} = \frac{\pi n}{214} \text{ kN mm}^{-2}$$

also from equation (iv)

$$K = 0.0112n$$

Substituting these values in equation (vi) we obtain the maximum stress in the bar as

$$\sigma_m = \frac{0.01716n}{1 + 0.0112n} \text{ kN mm}^{-2}$$

or $$\sigma_m = \frac{17.16n}{1 + 0.0112n} \text{ N mm}^{-2} \qquad (xi)$$

From equations (vii) and (x), the slenderness ratio of the bar is given by

$$\frac{L}{r} = 20\pi n \frac{d}{b}$$

After a number of trials it is found that a convenient value of the ratio d/b is 0.2 thus

$$\frac{L}{r} = 4\pi n \qquad (xii)$$

Now equation (xi) gives the actual maximum compressive stress in the bar, while the slenderness ratio obtained from equation (xii) can be used to determine allowable stresses ($p_c$) from Table 17c of BS 449. The results of some of the calculations for $\sigma_m$ and $P_c$ with varying n are shown in table 3.1.

Table 3.1

| n | $\sigma_m$ (N mm$^{-2}$) | L/r | $P_c$ (N mm$^{-2}$) |
|---|---|---|---|
| 7.0 | 111.4 | 88.0 | 123.0 |
| 7.1 | 112.9 | 89.2 | 120.4 |
| 7.2 | 114.3 | 90.5 | 117.0 |
| 7.3 | 115.8 | 91.7 | 114.6 |
| 7.4 | 117.3 | 93.0 | 112.0 |
| 7.5 | 118.7 | 94.2 | 109.6 |

The greatest value that n can have therefore lies between 7.2 and 7.3. Further calculation gives 7.26 as a better estimate. This value corresponds to an actual stress of 115.2 N mm$^{-2}$ and an allowable stress of 115.6 N mm$^{-2}$.

Substituting n = 7.26 and b = 5d in equation (x) we obtain

$\quad$ t = 1.14d $\quad$ or $\quad$ d = 0.877t

thus b = 4.38t

Substituting for b in equation (ix) we have

$\quad t^2 = 12.2 W_c$ mm$^2$

but $W_c$ = 200 kN, thus

$\quad$ t = 49.4 mm

hence

$\quad$ d = 43.3 mm

$\quad$ b = 216.4 mm

R = 494 mm

$F_0$ = 169.8 kN

and λ = 7.6 mm

The proportions of the ring and bar are now established for operation at a maximum screw-jack load of 200 kN. Since the tripping load is directly proportional to λ, the device can be operated at any load W, less than 200 kN, by setting the bar length to give

λ = 0.038W mm

if W is in kN.

The ring would be made by cutting a strip from plate of suitable thickness, forging into a ring and welding the ends. Final finishing to the required dimensions would be carried out on the lathe.

There are a number of refinements which could be added to improve the performance. For example, spring-loading the bar would permit the device to be used for vertical forces.

PROBLEMS

1. Derive expressions for the deflexions in a straight beam of span L and flexural rigidity EI resting on simple supports if the loading is: (a) a concentrated load W at mid-span; (b) a uniformly distributed load of intensity w per unit length.

Hence show that the mid-span deflexions are

$$\frac{WL^3}{48EI} \quad \text{and} \quad \frac{5wL^4}{384EI}$$

respectively.

2. A beam of uniform section rests on four simple supports A, B, C and D at the same level. AB = BC = CD = L. There is a load W at the middle of each span. Find the greatest bending moment in the beam assuming elastic behaviour.
[Cambridge]  [7WL/40]

3. A cantilever of span 2L and flexural rigidity EI carries a uniformly distributed load of intensity w per unit length over the whole span and a couple M at the free end. If the mid-point of the span is propped to the same level as the built-in end, show that the vertical deflexion at the free end is $wL^3/15EI$.

4. The davit shown in figure 3.23 is built-in at the foot and is made of solid circular section steel bar of diameter 100 mm. What is the greatest value that the load W can have if the maximum bending

stress in the davit is not to exceed 160 m$^{-2}$ and if the vertical deflexion at C is not to exceed 30 mm.  E = 200 GN m$^{-2}$.
[Sussex] [12.9 kN]

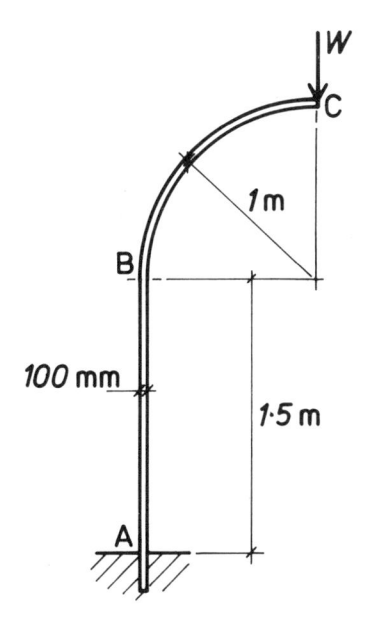

Figure 3.23

5.  The beam ABC of uniform section shown in figure 3.24 consists of a straight portion (AB) of length R and a quadrant (BC) of radius R. End A is built into a wall and end C is pinned to a roller support. Determine the maximum bending moment in the beam if a clockwise couple of 120 kN m is applied at B.
[Sussex] [67 kN m]

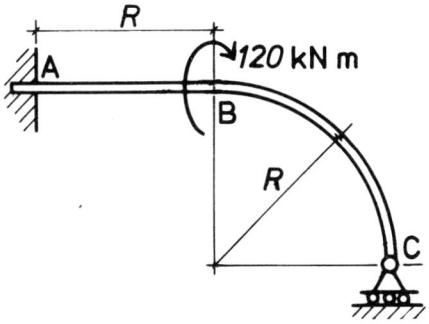

Figure 3.24

6. The frame ABCD shown in figure 3.25 is built-in at A and a horizontal force H is applied at D. The joints at B and C are rigid. AB = CD = h and BC = 1.6h. The second moments of area are $I_1$ for BC and $I_2$ for AB and CD. Show that if the path of point D under load is to make an angle of 45° with the horizontal, $I_1 = 2.4 I_2$.

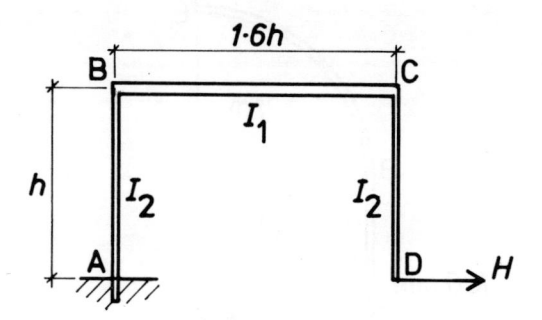

Figure 3.25

7. Figure 3.26 shows a flat ring fabricated from steel strip 25 mm wide and 8 mm thick. If the ring supports a load of 600 N as shown, determine the maximum bending moment and the relative deflexion of the load points.                                  [38.7 N m, 3.8 mm]

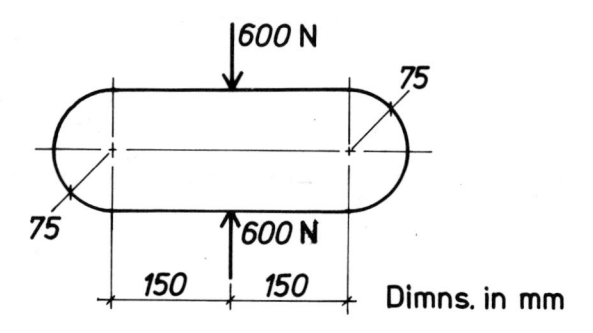

Figure 3.26

8. Figure 3.27 shows a ring of mean radius R made from a bar with the two ends at C connected by a frictionless pin-joint. The ring is subjected to three radial forces arranged in equilibrium as shown. show that the force on the pin is

$$\frac{2W}{3\pi}\left[8 + 2(\pi - \alpha)\right]$$

122

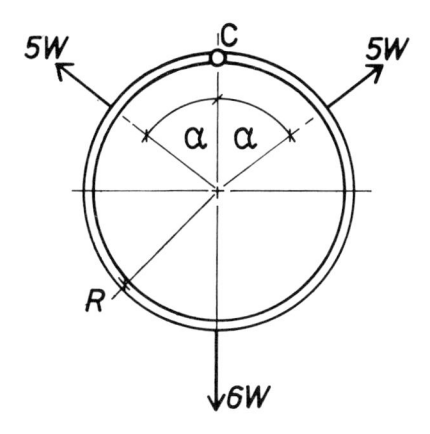

Figure 3.27

9. A semi-circular steel arch of mean radius R and flexural rigidity EI is shown in figure 3.28. The ends of the arch are built-in and a single concentrated vertical load W is carried at mid span. Determine the greatest bending moment in the arch and the deflexion at the load point. [0.15WR, $WR^3/86EI$]

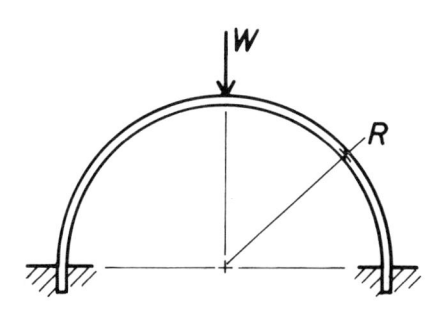

Figure 3.28

10. The parabolic arch shown in figure 3.29 is pinned at the footings A and B. With the origin at A (as shown) the equation of the parabola is

$$y = \frac{x}{20}(40 - x)$$

If the second moment of area of the arch cross-section varies directly as the secant of the slope of the arch, show that the

123

horizontal thrust, H due to a unit vertical load acting at a
distance x from A is given by

$$H = \frac{x}{48600}(27000 - 60x^2 + x^3)$$

[Sussex]

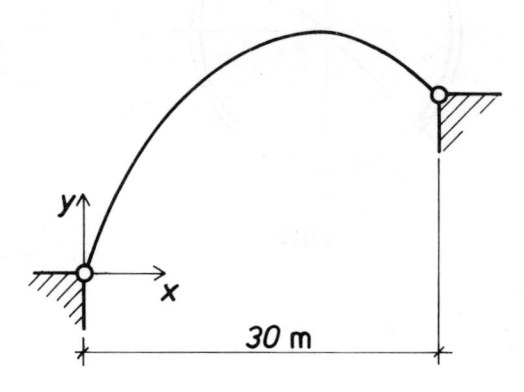

Figure 3.29

11. If end A of the spring in example 3.6 is built-in, determine the new stiffness. $[0.37EI/R^3]$

12. The link shown in figure 3.30 has cross-sectional dimensions which are small compared with R. Determine the maximum bending moment for the loading shown.
[Southampton] $[0.39WR]$

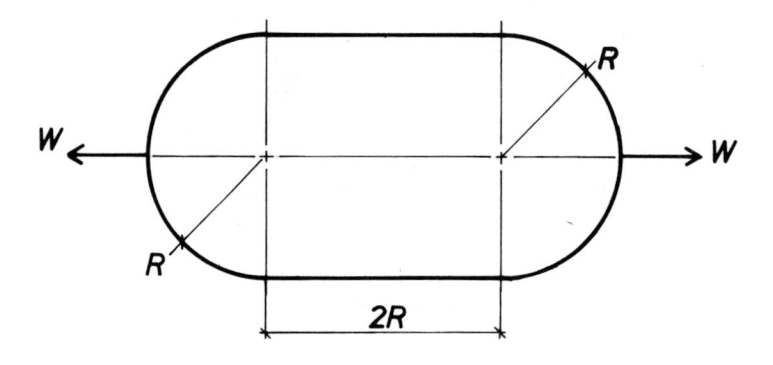

Figure 3.30

13. The planar structure shown in figure 3.31 consists of a semi-circular arch of radius 2 m which is supported through pinned connexions to two columns of height 4 m whose feet are built into the ground. Both the arch and the columns have the same in-plane flexural rigidity. If a concentrated vertical load of 10 kN is

applied to the top of the arch, determine the force and moment
reactions at the foot of each column.     [5 kN, 0.72 kN, 2.9 kN m]
[Sussex]

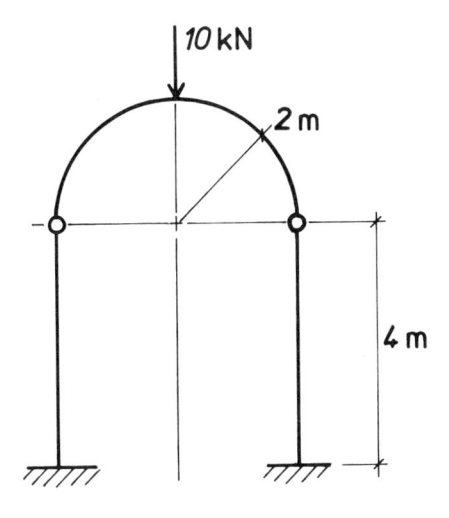

Figure 3.31

14. The rigidly jointed, plane rectangular loading frame shown in figure 3.32 is made of the same material throughout. The frame consists of two uprights of height 3 m having second moments of area I and two beams of length 5 m whose second moments of area are 2.5I. Show that the relative deflexion (Δ) of the load points A and B under the load W is given by

$$\Delta = \frac{35W}{24EI}$$

[Sussex]

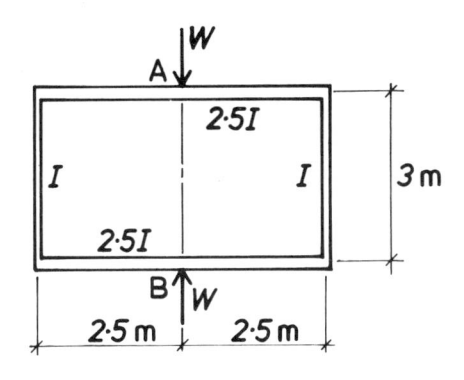

Figure 3.32

15. Figure 3.33 shows a square portal frame of height and span L. All members have the same uniform cross-section. The columns are both pin-jointed to the rigid base. A uniformly distributed horizontal load of w per unit length is applied to the left-hand column over its whole length. Find the position and value of the bending moment of greatest magnitude in the frame. Assume that the frame remains linear-elastic and that only deflexions due to bending need be considered. [Top right-hand joint, $M = 11wL^2/40$]
[Cambridge]

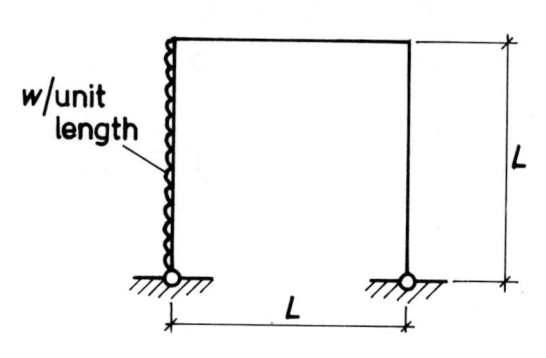

Figure 3.33

16. The frame ABCD shown in figure 3.34 is made of three different steel sections. The feet are pinned to the foundations at A and built-in at D. The beam, BC carries a uniformly distributed load of 3 kN m$^{-1}$ and a concentrated horizontal load of 12 kN is applied at joint B. Determine the greatest moment in the frame and the horizontal deflexion at B. [31.7 kN m, 132.3/EI]

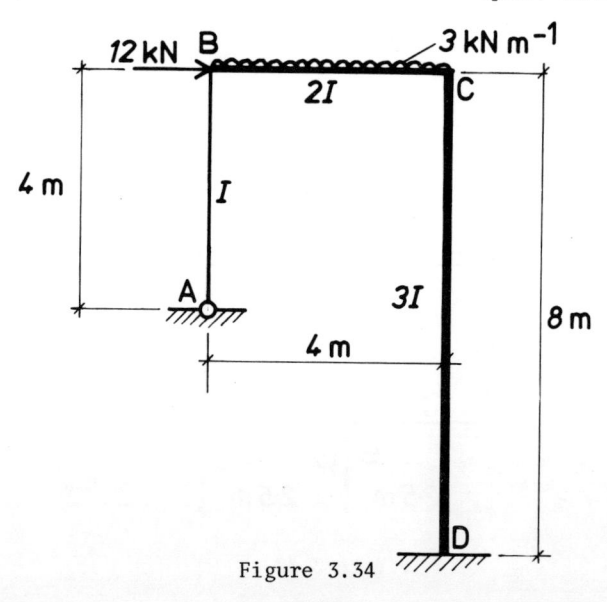

Figure 3.34

126

17. The plane frame ABCD shown in figure 3.35 has rigid joints at B and C and is pinned to the foundations at A and D. The horizontal member BC carries a uniformly distributed load of intensity 6 kN m$^{-1}$. Determine the maximum moment in the frame and the reactions at A and D.
[Sussex]      [24.7 kN m, 12.3 and 14.7 kN, 12.3 and 21.3 kN]

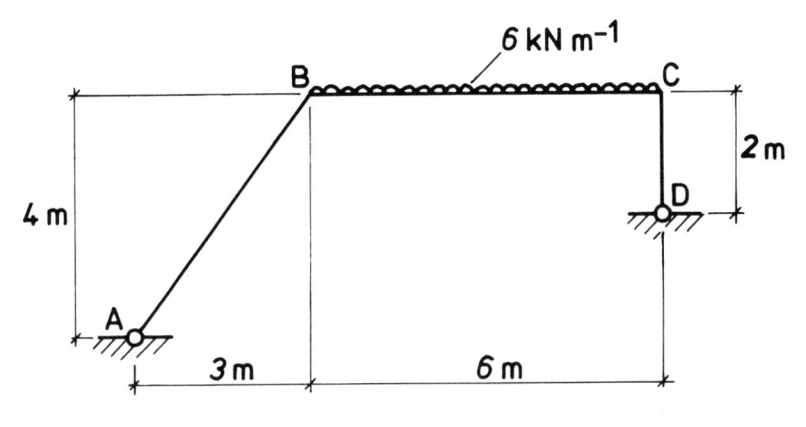

Figure 3.35

# 4 POTENTIAL ENERGY METHODS

The application of potential energy methods is primarily directed towards the determination of sets of unknown deflexions and, as such, is an equilibrium approach to structural analysis. We saw in section 1.3 that the fundamental theorem of potential energy is the principle of stationary potential energy which states that, for a structural system in equilibrium

$$\frac{\partial V}{\partial \Delta_j} = 0$$

(equation 1.5) where the potential energy V (defined in section 1.3) is expressed in terms of the displacements of the system.

We now need to inquire into the nature of this equilibrium state since the system may be stable or unstable. It will be shown in the next section that, in addition to a stationary value of the potential energy

(a) $\dfrac{\partial^2 V}{\partial \Delta_j^2}$ is >0 for stable equilibrium

(b) $\dfrac{\partial^2 V}{\partial \Delta_j^2}$ is <0 for unstable equilibrium

(c) if $\dfrac{\partial^2 V}{\partial \Delta_j^2} = 0$ the system may be stable or unstable

For a particular structural system, condition (a) represents a minimum in the function V while condition (b) is a maximum. Condition (c) merely indicates that a horizontal tangent exists, which may be the result of a minimum, a maximum or a point of inflexion.

To illustrate these concepts, the following section investigates a particular problem which exhibits both stable and unstable states of equilibrium.

## 4.1 CONDITIONS FOR EQUILIBRIUM: A CASE STUDY

In this section we shall consider the deflexion sensitive structure shown in figure 4.1. Two linearly elastic rods AB and BC are pinned together at B and to rigid abutments at A and C. Their combined unstressed length, 2L is slightly greater than the span 2a between the abutments. The rods are arranged in the vertical plane with B above the chord AC. Both rods have the same cross-sectional area (A) and modulus of elasticity (E). A vertical load W applied at B causes a vertical deflexion at B of $\Delta$. The initial height of B above AC is h. It will be noted that the deformation of the system is completely described by the deflexion Δ; we are therefore dealing with a system having only one degree of freedom.

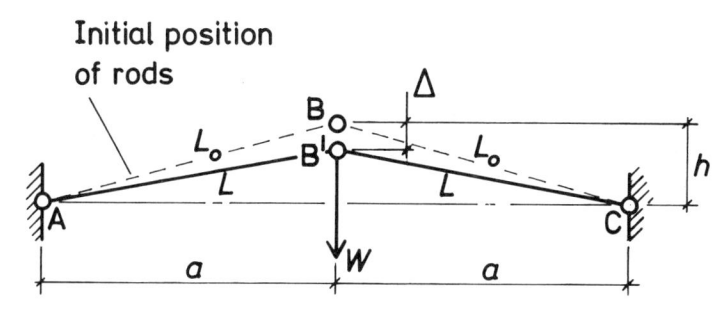

Figure 4.1

In the deflected position, the length of the rods is L, thus each rod is compressed by an amount δ given by

$$\delta = L_0 - L \tag{i}$$

where

$$L_0 = \sqrt{(h^2 + a^2)} \tag{ii}$$

and $L = \sqrt{[(h - \Delta)^2 + a^2]}$ (iii)

If h is small compared with a we have, from equations (i), (ii) and (iii)

$$\delta = \frac{h^2}{2a} \left(\frac{\Delta}{h}\right) \left(2 - \frac{\Delta}{h}\right) \tag{iv}$$

The strain energy (U) stored in both rods due to this deformation is given by

$$U = \frac{\delta^2 AE}{L_0} \tag{v}$$

From equations (ii), (iv) and (v) and again noting that h is small compared with a we find that

$$U = \frac{h^4 AE}{4a^3} \left(\frac{\Delta}{h}\right)^2 \left(2 - \frac{\Delta}{h}\right)^2 \tag{vi}$$

From the definition given in section 1.3 (equation (i)) the potential energy of the load (W) is given by

$$W_d = -W\Delta \tag{vii}$$

The potential energy (V) of the system is then obtained from equations (vi), (vii) and 1.4 as

$$V = \frac{h^4 AE}{4a^3}\left(\frac{\Delta}{h}\right)^2\left(2 - \frac{\Delta}{h}\right)^2 - W\Delta \tag{viii}$$

This equation is more conveniently expressed in the following non-dimensional form

$$\bar{V} = \frac{\bar{\Delta}^2}{4}(2 - \bar{\Delta})^2 - \bar{W}\bar{\Delta} \tag{ix}$$

where

$$\bar{V} = \frac{a^3 V}{h^4 AE}$$

$$\bar{W} = \frac{a^3 W}{h^3 AE}$$

and $\bar{\Delta} = \frac{\Delta}{h}$

For the system to be in equilibrium we require that

$$\frac{\partial \bar{V}}{\partial \bar{\Delta}} = \frac{1}{4}[2\bar{\Delta}(2 - \bar{\Delta})^2 - 2\bar{\Delta}^2(2 - \bar{\Delta})] - \bar{W} = 0$$

hence

$$\bar{W} = \bar{\Delta}(1 - \bar{\Delta})(2 - \bar{\Delta}) \tag{x}$$

also $\dfrac{\partial^2 \bar{V}}{\partial \bar{\Delta}^2} = \bar{V}'' = (2 - 6\bar{\Delta} + 3\bar{\Delta}^2)$ \hfill (xi)

The load-deflexion relationship of equation (x) is shown in figure 4.2. It can be seen that, although the system consists of linear elastic members, this relationship is non-linear in form. Also shown in figure 4.2 is the relationship, equation (xi), between the second differential of the potential energy function ($\bar{V}''$) and the deflexion.

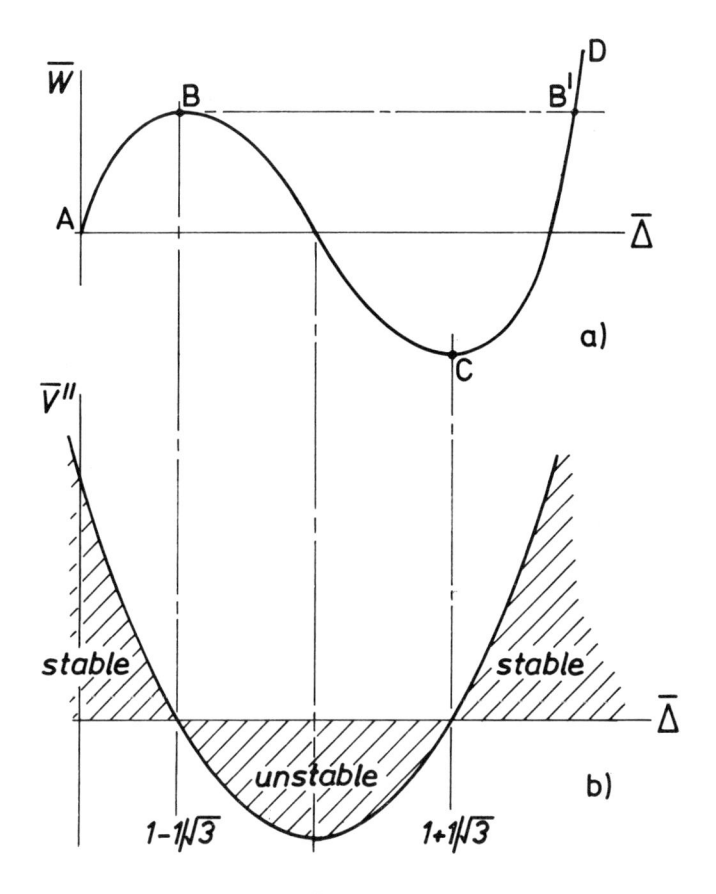

Figure 4.2

Examination of figure 4.2 shows that between A and B and C and D on the load-deflexion curve (figure 4.2a), equilibrium is stable because a positive value of $\bar{V}''$ denotes that $\bar{V}$ has a minimum for all corresponding values of $\bar{W}$ and $\bar{\Delta}$. Between B and C equilibrium is unstable because $\bar{V}''$ is negative and $\bar{V}$ has a maximum for all corresponding values of $\bar{W}$ and $\bar{\Delta}$. Precisely at B and C, $\bar{V}''$ is zero indicating that $\bar{V}$ has a minimax at these points.

To show that a minimum in the potential energy function corresponds to a stable condition of the structure or, conversely, that a maximum corresponds to an unstable condition, we need to look at the effect on $\bar{V}$ of small disturbances about the equilibrium position.

Suppose that the equilibrium position of the structure is defined by $\bar{W}$ and $\bar{\Delta}$ in accordance with equation (x) and that small changes in $\bar{\Delta}$ of magnitude $\eta$ are applied on each side of the equilibrium position. The change in the potential energy caused by these disturbances is therefore given by

$$\delta \bar{V} = \bar{V}(\bar{\Delta} + \eta) - \bar{V}(\bar{\Delta}) \tag{xii}$$

where η can be positive or negative.

A positive value of $\delta\bar{V}$ shows that energy must be added to the system to move it to an adjacent equilibrium state and thus the system is stable.

A negative value of $\delta\bar{V}$ indicates that a small disturbance in the equilibrium state results in a release of energy and thus the system must be unstable. The release of energy will continue until the system reaches a new, stable equilibrium position. It is assumed that the disturbances are not large enough to alter an inherently stable condition into one which is unstable.

After writing the first term on the right of equation (xii) in a Taylor's series and noting that $\bar{V}'(\bar{\Delta})$ is zero as a requirement for equilibrium, we have

$$\delta\bar{V} = \frac{\eta^2}{24}[12\bar{V}''(\bar{\Delta}) + 4\eta\bar{V}'''(\bar{\Delta}) + \eta^2\bar{V}''''(\bar{\Delta})]$$

For small values of η, the sign of $\delta\bar{V}$ is clearly dominated by the sign of $\bar{V}''(\bar{\Delta})$ thus confirming the assertion made earlier that positive values of $\bar{V}''(\bar{\Delta})$ indicate stable equilibrium states and vice versa.

When $\bar{V}''(\bar{\Delta})$ is zero, the change in potential energy is given by

$$\delta\bar{V} = \frac{\eta^3}{24}[4\bar{V}'''(\bar{\Delta}) + \eta\bar{V}''''(\bar{\Delta})]$$

For a particular value of $\bar{\Delta}$, the sign of $\delta\bar{V}$ will now depend on whether η is positive or negative, assuming $\bar{V}'''(\bar{\Delta}) \neq 0$. Thus a disturbance in one direction from the equilibrium position will require an input of energy to the system while a disturbance in the opposite direction will cause the system to jump to the next stable equilibrium state. Since the sign of η is arbitrary, this condition must be regarded as indicating an unstable equilibrium state. If $\bar{V}'''(\bar{\Delta}) = 0$, then stability will depend on the sign of $\bar{V}''''(\bar{\Delta})$ and so on.

As the load W on the structure in figure 4.1 is gradually increased from zero, the path AB on the load-deflexion curve of figure 4.2a is followed. At B, where $\bar{W} = 2/3\sqrt{3}$, the system becomes unstable and jumps to its next stable configuration at B'. This type of behaviour is often called 'snap-through' buckling; it is a particular feature of shell-type structures. The path BCB' can only be followed if the structure is tested in a device which is able to control the displacement.

A third equilibrium state can exist, but it is not exhibited in the structure we have been considering. Suppose that there is a system in which, under a certain loading condition, all the differentials of the energy function vanish identically, then $\delta V$ is zero and V is a constant. This condition is known as neutral equilibrium. An initially straight, pin-ended, axially loaded strut is in a condition of neutral equilibrium when subjected to the Euler buckling load. If the strut is displaced slightly from the straight configuration it will remain in its new position without showing any tendency to return to its original position or to seek out an alternative stable equilibrium state.

## 4.2 STRUCTURAL SYSTEMS WITH A LIMITED NUMBER OF DEGREES OF FREEDOM

There is no particular advantage in using the principle of stationary potential energy to obtain equilibrium equations for structures having only a few degrees of freedom unless, as in the previous section, we wish to examine the stability of the system. In chapter 1, Castigliano's first theorem, part I, was used to show that a stationary value of the potential energy corresponded with an equilibrium state. Thus if exact solutions to a particular problem are required, the two approaches are exactly equivalent.

For example, Castigliano's first theorem can be applied to obtain equation (x) of section 4.1. Differentiating the strain energy expression (equation vi) with respect to $\Delta$ and equating the result to the load W gives equation (x) directly.

Alternatively the equilibrium equations for the various examples in section 2.5 could have been obtained by application of the principle of stationary potential energy. As an illustration, the potential energy expression for the pin-jointed frame of example 2.11 is

$$V = \sum_{1}^{4} \frac{EA}{2L} \Delta_T^2 - 10\Delta_h - 5\Delta_V \tag{i}$$

For equilibrium we require that

$$\frac{\partial V}{\partial \Delta_H} = \frac{\partial V}{\partial \Delta_V} = 0$$

hence

$$\sum_{1}^{4} \frac{EA}{L} \Delta_T \frac{\partial \Delta_T}{\partial \Delta_H} - 10 = 0$$

and

$$\sum_{1}^{4} \frac{EA}{L} \Delta_T \frac{\partial \Delta_T}{\partial \Delta_V} - 5 = 0$$

as before.

It is easy to show that both equilibrium equations describe stable conditions, for if we differentiate equation (i) twice with respect to $\Delta_H$ and $\Delta_V$, we have

$$\frac{\partial^2 V}{\partial \Delta_H^2} = \sum_{1}^{4} \frac{EA}{L} \left[ \left(\frac{\partial \Delta_T}{\partial \Delta_H}\right)^2 + \Delta_T \frac{\partial^2 \Delta_T}{\partial \Delta_H^2} \right]$$

and $\dfrac{\partial^2 V}{\partial \Delta_V^2} = \sum\limits_1^4 \dfrac{EA}{L}\left[\left(\dfrac{\partial \Delta_T}{\partial \Delta_V}\right)^2 + \Delta_T \dfrac{\partial^2 \Delta_T}{\partial \Delta_V^2}\right]$

The $\Delta_T$s are linear functions of $\Delta_H$ and $\Delta_V$, thus the second terms in the brackets are zero. Since the first terms are always positive, $V''$ must also be positive.

If we attempt to use the principle of stationary potential energy to obtain exact equilibrium equations for a system having many degrees of freedom we will quickly find that the labour of solving the corresponding simultaneous equations makes the approach impracticable. If the number òf degrees of freedom are infinite, an exact solution by this method is impossible and we must turn to the approximate procedure discussed in the next section.

## 4.3 APPROXIMATE SOLUTIONS: THE RAYLEIGH-RITZ METHOD

Certain problems in engineering are extremely difficult, if not impossible, to solve exactly. A relatively simple procedure for dealing with such intractable problems was devised originally by the British scientist Lord Rayleigh (1842-1919) and later extended by the Swiss mathematician Walter Ritz.

The basic approach is to assume a suitable displacement or shape function for the elastic system under consideration. This function may contain one or more unknown coefficients which can be determined by application of the principle of stationary potential energy. Strictly speaking, the use of a displacement function with one unknown coefficient is referred to as the Rayleigh method. The contribution of Ritz was to introduce an infinite series for the displacement function thus permitting any desired improvement in the accuracy of a solution simply by considering a sufficient number of terms in the series.

Although the Rayleigh-Ritz method can be applied to any linear or non-linear elastic structural system, attention in the following will be confined to beams and columns of linear elastic material.

*4.3.1 The Treatment of Beams*

The only restriction on the choice of a displacement function is that it must satisfy the geometric boundary conditions of the problem. However, the solution is likely to be much more accurate if the boundary curvature conditions are also satisfied. In some cases it might also be necessary to examine the third derivative. The following example shows how a poor choice for the displacement function can lead to a solution being found for an entirely different problem.

*Example 4.1*

A simply supported linear-elastic beam of span L carries a

concentrated load (W) at mid-span.  Assume a suitable function to
define the deflected shape and, by minimising the potential energy,
estimate the mid-span deflexion.

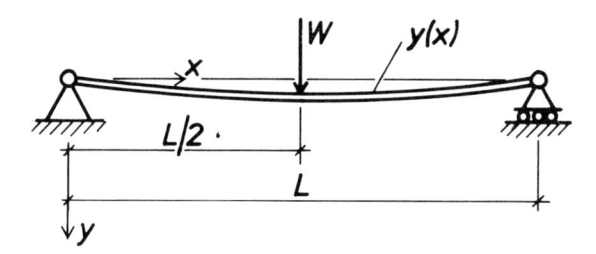

Figure 4.3

Figure 4.3 shows the beam and the assumed deflected shape.
Suppose we choose a quadratic as the displacement function, then

$$y(x) = a_0 + a_1 x + a_2 x^2$$

where y is the beam deflexion at distance x from one support.

We must at least satisfy the geometric boundary conditions $y = 0$
at $x = 0$ and L, thus

$$a_0 = 0 \quad \text{and} \quad a_2 = - \frac{a_1}{L}$$

hence

$$y(x) = a_1 x (1 - \frac{x}{L}) \qquad (i)$$

The function $y(x)$ now contains one unknown coefficient which may be
determined by minimising the potential energy.

From section 3.1 we have an expression for the complementary
energy due to bending.  Since this was derived for a linear-elastic
beam we may use the same expression for the strain energy, thus

$$U = \int \frac{M^2}{2EI} dx \qquad (ii)$$

but, also from section 3.1

$$M = \frac{EI}{R} \qquad (iii)$$

where, for small deflexions

$$\frac{1}{R} = \frac{d^2 y}{dx^2} = y'' \qquad (iv)$$

Thus, substituting for M in equation (ii) we have

$$U = \frac{EI}{2}(y'')^2 \, dx \tag{v}$$

From equation (i), $y'' = -2a_1/L$ and

$$U = \int_0^L \frac{2EIa_1^2}{L^2} \, dx = \frac{2EIa_1^2}{L}$$

The work done by the load W is given by

$$W_d = -Wy\frac{L}{2} = -W\frac{a_1 L}{4}$$

hence the total potential energy of the system is

$$V = \frac{2EIa_1^2}{L} - W\frac{a_1 L}{4}$$

For equilibrium, since $a_1$ is the only variable

$$\frac{\partial V}{\partial a_1} = \frac{2EIa_1}{L} - \frac{WL}{4} = 0$$

from which

$$a_1 = \frac{WL^2}{16EI}$$

and the mid-span deflexion is therefore

$$y\left(\frac{L}{2}\right) = \frac{a_1 L}{4} = \frac{WL^3}{64EI} \tag{vi}$$

We know that the correct value (see *Essential Solid Mechanics*) of this deflexion should be $WL^3/48EI$, thus the result given by equation (vi) is by no means accurate. The reason for this is that although the displacement function chosen satisfies the geometric boundary conditions, an important statical boundary condition (that the curvature is zero at x = 0 and L) is violated.

The bending moments in the beam which we are supposed to be investigating are zero at the ends and increase linearly to a value of WL/4 at mid-span. However, for the beam represented by the displacement function of equation (i), the bending moment is constant, since from equations (i), (iii) and (iv)

$$M = EIy'' = -\frac{2EIa_1}{L}$$

where

$$a_1 = \frac{WL^2}{16EI}$$

thus $M = -\frac{WL}{8}$

The deflexion given by equation (vi) is therefore more likely to be an accurate result for a beam of span L having a constant bending moment produced by a pair of equal and opposite couples of magnitude -WL/8 acting at its ends. The negative sign denotes that the couples are producing a sagging moment in the beam. As it happens, the deflexion given by equation (vi) is exact because equation (i) is a correct representation of the shape of a beam subject to constant bending moment.

A much better choice of displacement function for the beam of figure 4.3 would have been

$$y(x) = a \sin \frac{\pi x}{L}$$

from which

$$y'' = -\frac{\pi^2 a}{L^2} \sin \frac{\pi x}{L}$$

thus both the geometric boundary conditions ($y = 0$ at $x = 0$ and $L$) and the statical boundary conditions ($y'' = 0$ at $x = 0$ and $L$) are now satisfied.

Proceeding as before, we have

$$U = \int_0^L \frac{EI}{2} \frac{\pi^4 a^2}{L^4} \sin^2 \frac{\pi x}{L} \, dx$$

and $W_d = -Wa$

thus $V = \frac{\pi^4 EI a^2}{4L^3} - Wa$

For equilibrium, since a is the only variable

$$\frac{\partial V}{\partial a} = \frac{\pi^4 EI a}{2L^3} - W = 0$$

hence a (the mid-span deflection) is given by

$$a = \frac{2WL^3}{\pi^4 EI} = \frac{WL^3}{48.7 EI}$$

This result is within 1.5% of the correct answer which we know to be $WL^3/48EI$.

Although the mid-span deflexion obtained above is a great improvement on that found previously (equation vi) we can do better still by assuming a displacement function in the form of an infinite series. For example, if the deflected shape of the beam is given by the odd terms of a sine series we have

$$y(x) = \sum_{1,3,5...}^{\infty} a_n \sin \frac{n\pi x}{L} \qquad \text{(vii)}$$

The even terms are not wanted because they are asymmetric with respect to the mid-span of the beam.

From equation (vii)

$$y'' = -\frac{\pi^2}{L^2} \sum_{1,3,5...}^{\infty} n^2 a_n \sin \frac{n\pi x}{L} \qquad \text{(viii)}$$

The required geometric and statical boundary conditions are therefore satisfied by $y$ and $y''$ from equations (vii) and (viii) and the strain energy is given by

$$U = \frac{\pi^4 EI}{2L^4} \int_0^L \left( \sum_{1,3,5...}^{\infty} n^2 a_n \sin \frac{n\pi x}{L} \right)^2 dx$$

since $EI$ is constant.

Although the integral appears to be somewhat daunting, it should be noted that squaring the series produces terms of two possible forms, either

$$n^4 a_n^2 \sin^2 \frac{n\pi x}{L}$$

or $\quad n^2 m^2 a_n a_m \sin \frac{n\pi x}{L} \sin \frac{m\pi x}{L}, \quad n \neq m$

Fortunately sine functions (in common with cosine functions and certain special polynomials) have the property of orthogonality which means that the integrals of all the mixed product terms are zero. Also

$$\int_0^L \sin^2 \frac{n\pi x}{L} dx = \frac{L}{2}$$

hence the expression for $U$ may now be written as

$$U = \frac{\pi^4 EI}{4L^3} \sum_{1,3,5...}^{\infty} n^4 a_n^2$$

The potential energy of the load (at $x = L/2$) is given by

$$W_d = -W \sum_{1,3,5...}^{\infty} a_n \sin \frac{n\pi}{2}$$

The total potential energy is therefore

$$V = \frac{\pi^4 EI}{4L^3} \sum_{1,3,5...}^{\infty} n^4 a_n^2 - W \sum_{1,3,5...}^{\infty} a_n \sin \frac{n\pi}{2}$$

There are an infinite number of coefficients ($a_n$) describing the deflected form of the beam. The potential energy (V) must have a stationary value with respect to each one, therefore in general

$$\frac{\partial V}{\partial a_n} = \frac{\pi^4 EI}{4L^3} n^4 \times 2a_n - W \sin \frac{n\pi}{2} = 0$$

or $\quad a_n = \dfrac{2WL^3}{\pi^4 EI} \dfrac{1}{n^4} \sin \dfrac{n\pi}{2}$

hence

$$y(x) = \frac{2WL^3}{\pi^4 EI} \sum_{1,3,5...}^{\infty} \frac{1}{n^4} \sin \frac{n\pi}{2} \sin \frac{n\pi x}{L} \qquad (ix)$$

This expression is exact since an infinite number of terms are involved which correspond to the infinite number of degrees of freedom in the system.

The mid-span deflexion is obtained by putting $x = L/2$ in equation (ix) thus

$$y\left(\frac{L}{2}\right) = \frac{2WL^3}{\pi^4 EI} \sum_{1,3,5...}^{\infty} \frac{1}{n^4} \sin^2 \frac{n\pi}{2}$$

but, since n is odd, $\sin^2 n\pi/2 = 1$, therefore

$$y\left(\frac{L}{2}\right) = \frac{2}{\pi^4} \frac{WL^3}{EI} \sum_{1,3,5...}^{\infty} \frac{1}{n^4}$$

or $\quad y\left(\dfrac{L}{2}\right) = \dfrac{2}{\pi^4} \dfrac{WL^3}{EI} \left(\dfrac{1}{1^4} + \dfrac{1}{3^4} + \dfrac{1}{5^4} + \dfrac{1}{7^4} + ...\right)$

The improvement in the estimate of the mid-span deflexion obtained by taking more and more terms of the series can be seen in table 4.1. Remember that the exact value of $EIy(L/2)/WL^3$ is $1/48$. An interesting conclusion from this result is in the evaluation of $\pi$ from the infinite series $\left(96 \sum_{1,3,5...}^{\infty} n^{-4}\right)^{\frac{1}{4}}$.

Table 4.1

| Number of terms in series | 1 | 2 | 3 | 4 | 5 |
|---|---|---|---|---|---|
| $\dfrac{EI}{WL^3} y\left(\dfrac{L}{2}\right)$ | $\dfrac{1}{48.70}$ | $\dfrac{1}{48.11}$ | $\dfrac{1}{48.035}$ | $\dfrac{1}{48.015}$ | $\dfrac{1}{48.008}$ |

The Rayleigh-Ritz method is applicable to statically indeterminate problems without any further complications. No compatibility equations for the redundancies are needed, all that is required is that the displacement function should satisfy as many of the boundary conditions as possible. The next example deals with a statically indeterminate problem and introduces the procedure for determining the work done by a uniformly distributed load.

*Example 4.2*

A cantilever AB of span L is built in to a wall at A and carries a uniformly distributed load of intensity w per unit length over the whole span. If end B is propped to the same level as A, determine the force in the prop by minimising the total potential energy. Assume a quartic polynomial for the displacement function.

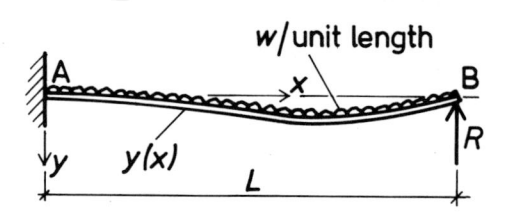

Figure 4.4

Figure 4.4 shows the propped cantilever and the assumed deflected shape. A general quartic in x has five coefficients, four of which may be obtained by satisfying boundary conditions. The remaining coefficient is found by minimising the total potential energy and hence establishing equilibrium conditions. For an origin at A, the displacement function may be written as

$$y(x) = a_0 + a_1 x + a_2 x^2 + a_3 x^3 + a_4 x^4 \qquad (i)$$

The geometric boundary conditions to be satisfied are $y = 0$ at $x = 0$ and L and $y' = 0$ at $x = 0$. The only statical boundary condition known is $y'' = 0$ at $x = L$. If these conditions are introduced into equation (i) we obtain

$$a_0 = 0$$

$$a_1 = 0$$

$$a_2 = \frac{3a_4}{2} L^2$$

and $\quad a_3 = -\frac{5a_4}{2} L$

thus $\quad y(x) = \frac{a_4 L^4}{2} \left(\frac{x}{L}\right)^2 \left[3 - 5\left(\frac{x}{L}\right) + 2\left(\frac{x}{L}\right)^2\right]$ \hfill (ii)

The strain energy is given by

$$U = \int_0^L \frac{EI}{2}(y'')^2 \, dx$$

and from equation (ii)

$$y'' = a_4 L^2 \left[3 - 15\left(\frac{x}{L}\right) + 12\left(\frac{x}{L}\right)^2\right] \hfill \text{(iii)}$$

hence

$$U = \frac{9EI}{10} a_4^2 L^5$$

The potential energy of a small element of load is $w \, dx \, y$, thus the potential energy of the whole of the distributed load is

$$W_d = -\int_0^L wy \, dx$$

thus

$$W_d = -\frac{3wa_4 L^5}{40}$$

and the total potential energy becomes

$$V = \frac{9EI}{10} a_4^2 L^5 - \frac{3wa_4}{40} L^5$$

To satisfy the requirement of equilibrium we have

$$\frac{\partial V}{\partial a_4} = \frac{9EI}{5} a_4 L^5 - \frac{3wL^5}{40} = 0$$

or $\quad a_4 = \frac{w}{24EI}$

141

In this problem we are interested in forces rather than deflexions. To determine the prop reaction R we note that the moment at A is given by

$$M_A = EI y''(0)$$

Substituting for $a_4$ in equation (iii) and putting $x = 0$ we have

$$y''(0) = \frac{wL^2}{8EI}$$

thus $M_A = \dfrac{wL^2}{8}$

but from figure 4.2 we see that

$$M_A = -RL + \frac{wL^2}{2}$$

thus $\dfrac{wL^2}{8} = -RL + \dfrac{wL^2}{2}$

or $R = \dfrac{3wL}{8}$

As we have seen in example 3.3, this is the correct result since a quartic polynomial in x completely describes the displacement of a uniformly loaded, propped cantilever.

*Example 4.3*

The non-prismatic beam shown in figure 4.5 is built in at the ends and carries a single concentrated load W at mid-span. On the assumption that a quartic polynomial in x describes the deflected shape of the beam, estimate the mid-span deflexion and the end moments.

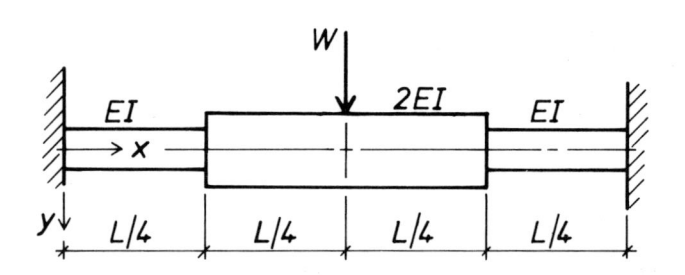

Figure 4.5

The displacement function to be assumed is
$$y(x) = a_0 + a_1 x + a_2 x^2 + a_3 x^3 + a_4 x^4$$
and the boundary conditions to be satisfied are

$y = 0$, $x = 0$ and $L$

$y' = 0$, $x = 0$ and $L$

thus

$a_0 = a_1 = 0$

$a_2 = a_4 L^2$

and $a_3 = -2 a_4 L$

Let the mid-span deflexion be $\Delta$, then

$$a_4 = \frac{16\Delta}{L^4}$$

and $y(x) = 16\Delta \left(\frac{x}{L}\right)^2 \left[1 - 2\left(\frac{x}{L}\right) + \left(\frac{x}{L}\right)^2\right]$

hence

$$y'' = \frac{32\Delta}{L^2}\left[1 - 6\left(\frac{x}{L}\right) + 6\left(\frac{x}{L}\right)^2\right]$$

Since the beam is made up of three sections of constant flexural rigidity, the total strain energy is determined from

$$U = \frac{EI}{2} \int_0^{L/4} (y'')^2 \, dx + \frac{2EI}{2} \int_{L/4}^{3L/4} (y'')^2 \, dx + \frac{EI}{2} \int_{3L/4}^{L} (y'')^2 \, dx$$

Now $(y'')^2 = \frac{1024\Delta^2}{L^4}\left[1 - 12\left(\frac{x}{L}\right) + 48\left(\frac{x}{L}\right)^2 - 72\left(\frac{x}{L}\right)^3 + 36\left(\frac{x}{L}\right)^4\right]$

hence

$$U = \frac{708 EI \Delta^2}{5L^3} \qquad \text{(i)}$$

The potential energy of the load is given by

$$W_d = -W\Delta \qquad \text{(ii)}$$

therefore the total potential energy is obtained from equations (i) and (ii) as

$$V = \frac{708EI\Delta^2}{5L^3} - W\Delta$$

For equilibrium

$$\frac{\partial V}{\partial \Delta} = \frac{1416EI\Delta}{5L^3} - W = 0$$

thus $\Delta = \frac{5WL^3}{1416EI} = 0.00353 \frac{WL^3}{EI}$

The exact answer is $\Delta = 0.00358WL^3/EI$ so that the energy solution is in error by only about 1.4%.

To determine the principle bending moments in the beam we note that

$$M = (EI)_x \, y'' \qquad \qquad (iii)$$

where $(EI)_x$ represents the flexural rigidity at x from the left-hand end (see figure 4.3); thus

$$M = (EI)_x \frac{20WL}{177EI}\left[1 - 6\left(\frac{x}{L}\right) + 6\left(\frac{x}{L}\right)^2\right]$$

At the ends of the beam, $x = 0$ and $L$ and $(EI)_x = EI$, hence

$$M_{x=0} = M_{x=L} = \frac{20WL}{177} = 0.113WL$$

At mid-span, $x = L/2$ and $(EI)_x = 2EI$, thus

$$M_{x=L/2} = -2\frac{10WL}{177} = -0.113WL$$

The exact results are

$$M_{x=0,L} = \frac{5WL}{48} = 0.104WL$$

and $M_{x=L/2} = -\frac{7WL}{48} = -0.146WL$

The approximate results for the moments are thus much less accurate than those obtained for the deflexions. This is generally so for the Rayleigh-Ritz method since the moments are obtained by differentiating the displacement function twice. If the displacement function is approximate, the second differential will be even less exact.

In this problem the error in the bending moment obtained from equation (iii) is particularly bad near mid-span because the second differential of the displacement function does not exhibit the discontinuity which must occur under the concentrated load. Figure

4.6 shows a comparison between the approximate and exact bending moment diagrams for the beam of figure 4.5.

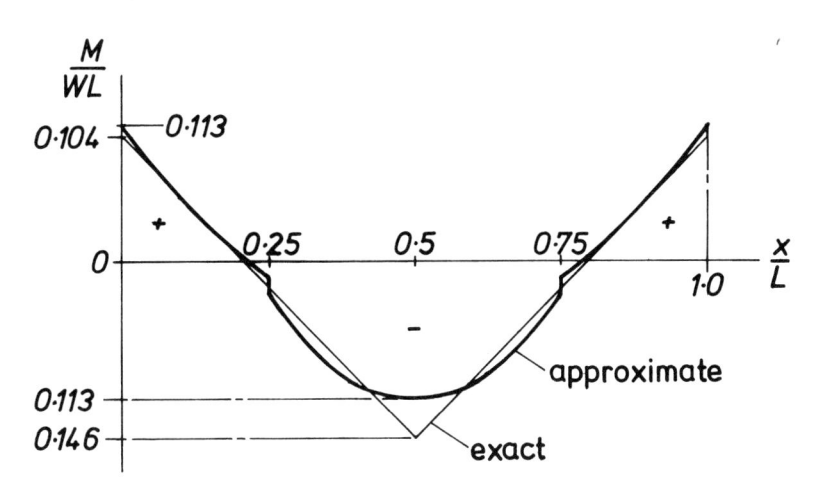

Figure 4.6

*Example 4.4*

A linearly elastic beam of non-uniform cross-section and span L is simply supported at each end. The beam carries a uniformly distributed vertical load of w per unit length. The beam has a horizontal centroidal axis and the second moment of area (I) for bending in the vertical plane is given by

$$I = I_0(1 + \pi \sin \frac{\pi x}{L})$$

the origin of the co-ordinate axes being at the left-hand support.

If the deflected shape of the beam is assumed to be given approximately by

$$y(x) = \Delta \sin \frac{\pi x}{L}$$

show, by minimising the potential energy, that

$$\Delta = \frac{12}{11\pi^5} \frac{wL^4}{EI_0}$$

Since $y(x) = \Delta \sin (\pi x/L)$

$$y'' = -\frac{\pi^2 \Delta}{L^2} \sin \frac{\pi x}{L} \tag{i}$$

The strain energy due to bending is given by

$$U = \frac{E}{2}\int_0^L I(y'')^2\,dx$$

where the second moment of area, I is included under the integral since it is a function of x.

Substituting for I, we obtain

$$U = \frac{\pi^4 EI_0 \Delta^2}{2L^4}\int_0^L \left(1 + \pi \sin\frac{\pi x}{L}\right)\sin^2\frac{\pi x}{L}\,dx$$

but $\int_0^L \sin^2\frac{\pi x}{L}\,dx = \frac{L}{2}$

and $\int_0^L \sin^3\frac{\pi x}{L}\,dx = \frac{4L}{3\pi}$

thus $U = \frac{11\pi^4}{12L^3}EI_0\Delta^2$

The potential energy of the load is given by

$$W_d = -\int_0^L wy\,dx = -\frac{2wL\Delta}{\pi}$$

thus the total potential energy is

$$V = \frac{11\pi^4}{12L^3}EI_0\Delta^2 - \frac{2wL\Delta}{\pi}$$

For equilibrium

$$\frac{\partial V}{\partial \Delta} = \frac{11\pi^4}{12L^3}EI_0 2\Delta - \frac{2wL}{\pi} = 0$$

from which

$$\Delta = \frac{12}{11\pi^5}\left(\frac{wL^4}{EI_0}\right)$$

*Example 4.5*

The tee-section cantilever shown in figure 4.7 has a span of 4 m and is cut from a 914 × 305 × 253 Universal Beam (UB). The overall depth varies linearly from 306 mm at the free end to 612 mm at the built-in end. A concentrated load of 65 kN is carried at the free end and an anticlockwise couple of 120 kN m is applied at mid-span.

A good approximation of the major axis second moment of area of the tee-section is given by

$$I = I_0(1 + ax + bx^2) \qquad 0<x<4 \text{ m}$$

where $I_0 = 9742 \text{ cm}^4$, $a = 0.685 \text{ m}^{-1}$, and $b = 0.210 \text{ m}^{-2}$.

Estimate the maximum deflexion $\Delta$ on the assumption that a suitable displacement function is

$$y = \Delta(1 - \sin \frac{\pi x}{8}) \qquad 0<x<4 \text{ m}$$

In the expressions for I and y, x (in m) is measured from the free end. $E = 200 \text{ GN m}^{-2}$.

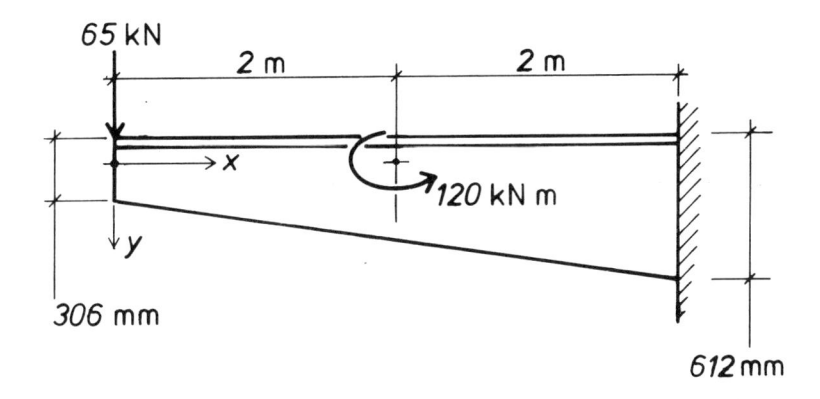

Figure 4.7

From the given displacement function we have

$$y' = -\frac{\pi \Delta}{8} \cos \frac{\pi x}{8} \qquad (i)$$

and $$y'' = \frac{\pi^2 \Delta}{64} \sin \frac{\pi x}{8} \qquad (ii)$$

Since the cantilever is of non-uniform section, the bending strain energy is given by

$$U = \frac{E}{2} \int_0^4 I(y'')^2 \, dx$$

147

Substituting for I from the data given and for $y''$ from equation (ii) we have

$$U = \frac{\pi^4 EI_0 \Delta^2}{8192} \int_0^4 (1 + ax^2 + bx^2) \sin^2 \frac{\pi x}{8} \, dx$$

To evaluate the integral it is helpful to make the substitution $\pi x/8 = \theta$, thus

$$U = \frac{\pi^3 EI_0 \Delta^2}{1024} \int_0^{\pi/2} \left(\sin^2 \theta + \frac{8a}{\pi} \theta \sin^2 \theta + \frac{64 b^2}{\pi^2} \theta^2 \sin^2 \theta\right) d\theta$$

but 
$$\int_0^{\pi/2} \sin^2 \theta \, d\theta = \left[\frac{\theta}{2} - \frac{\sin 2\theta}{4}\right]_0^{\pi/2} = \frac{\pi}{4}$$

$$\int_0^{\pi/2} \theta \sin^2 \theta \, d\theta = \left[\frac{\theta^2}{4} - \frac{\theta \sin 2\theta}{4} - \frac{\cos 2\theta}{8}\right]_0^{\pi/2} = \frac{\pi^2 + 4}{16}$$

and 
$$\int_0^{\pi/2} \theta^2 \sin^2 \theta \, d\theta = \left[\frac{\theta^3}{6} - \left(\frac{\theta^2}{4}\right) - \frac{1}{8} \sin 2\theta - \theta \frac{\cos 2\theta}{4}\right]_0^{\pi/2} = \frac{(\pi^2 + 6)}{\pi 48}$$

hence

$$U = \frac{\pi^3 EI_0 \Delta^2}{1024} \left[\frac{\pi}{12}(3 + 6a + 16b) + \frac{2}{\pi}(a + 4b)\right]$$

Inserting values for a and b and noting that $EI_0 = 19484$ kN m² we obtain

$$U = 2190 \Delta^2 \text{ kN m, } \Delta \text{ in m} \qquad (iii)$$

The potential energy of the loads is given by

$$W_d = -65\Delta - 120 \frac{\pi \Delta}{8\sqrt{2}} \text{ kN m, } \Delta \text{ in m}$$

where the second term represents the work done by the couple in rotating through an angle corresponding to the slope of the cantilever at $x = 2$ m. Simplifying, we obtain

$$W_d = -98.32\Delta \text{ kN m, } \Delta \text{ in m} \qquad (iv)$$

From equations (iii) and (iv), the total potential energy is given by

$$V = 2190\Delta^2 - 98.32\Delta \text{ kN m, } \Delta \text{ in m}$$

For equilibrium

$$\frac{\partial V}{\partial \Delta} = 4380\Delta - 98.32 = 0$$

or  $\Delta = 22.4 \times 10^{-3}$ m = 22.4 mm

### 4.3.2 The Effect of Axial Loads

A member which carries both axial and transverse loads is known as a beam-column.

When deriving the total potential energy in a beam-column, the strain energy due to direct stress is ignored since it is usually very small compared with the bending strain energy induced by the transverse loads. The expression for the total potential energy is therefore the same as that for a beam except that an additional term is required to account for the potential energy of the axial load. The following examples show how this term is obtained.

*Example 4.6*

The simply supported beam-column of span L and uniform cross-section shown in figure 4.8a is subjected to an axial load (P) together with a concentrated transverse (W) applied at distance c from the left-hand end. By minimising the total potential energy, determine the deflexion under the transverse load.

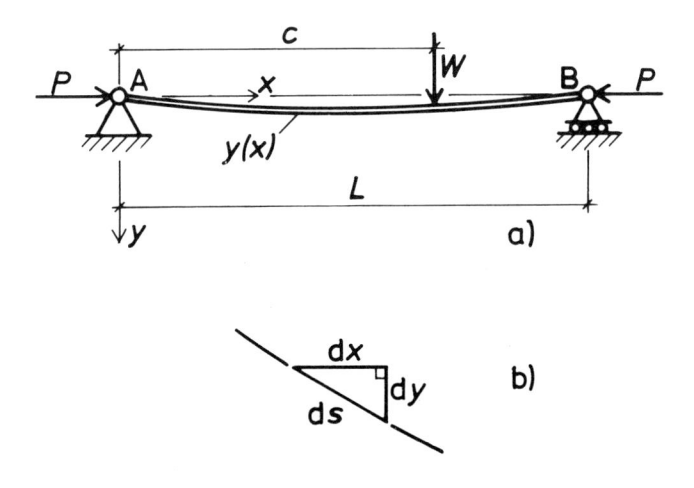

Figure 4.8

A suitable displacement function would be the infinite sine series given by

$$y(x) = \sum_{1}^{\infty} a_n \sin \frac{n\pi x}{L} \qquad \text{(i)}$$

from which

$$y' = \frac{\pi}{L} \sum_{1}^{\infty} n a_n \cos \frac{n\pi x}{L} \qquad (ii)$$

and
$$y'' = -\frac{\pi^2}{L^2} \sum_{1}^{\infty} n^2 a_n \sin \frac{n\pi x}{L} \qquad (iii)$$

The strain energy due to bending is thus

$$U = \frac{\pi^4 EI}{2L^4} \int_0^L \left( \sum_{1}^{\infty} n^2 a_n \sin \frac{n\pi x}{L} \right)^2 dx$$

or (see example 4.1)

$$U = \frac{\pi^4 EI}{4L^3} \sum_{1}^{\infty} n^4 a_n^2 \qquad (iv)$$

The potential energy of the loads is given by

$$W_d = -W \left( \sum_{1}^{\infty} a_n \sin \frac{n\pi c}{L} \right) - Ph$$

where h is the distance moved by the axial load P.

The distance h is equal to the difference between the beam length measured round the arc and the chord length AB. Referring to figure 4.8b we have

$$h = \int_0^L (ds - dx)$$

but $ds^2 = dx^2 + dy^2$, thus

$$h = \int_0^L \left\{ \sqrt{[1 + (y')^2]} - 1 \right\} dx$$

where $y' = dy/dx$.

The square root term may be expanded in a binomial series. For small deflexions, powers of $y'$ greater than the second may be neglected, thus

$$h = \tfrac{1}{2} \int_0^L (y')^2 \, dx \qquad (v)$$

From equations (ii) and (v) we have

$$h = \frac{\pi^2}{2L^2} \int_0^L \left( \sum_1^\infty n a_n \cos \frac{n\pi x}{L} \right)^2 dx$$

Like the sine series, the cosine series possesses the property of orthogonality, hence

$$\int_0^L \cos^2 \frac{n\pi x}{L} dx = \frac{L}{2}$$

and $\int_0^L \cos \frac{n\pi x}{L} \cos \frac{m\pi x}{L} dx = 0, \; n \neq m$

thus the final expression for $h$ is

$$h = \frac{\pi^2}{4L} \sum_1^\infty n^2 a_n^2$$

and the potential energy of the loads becomes

$$W_d = -W \sum_1^\infty a_n \sin \frac{n\pi c}{L} - \frac{\pi^2 P}{4L} \sum_1^\infty n^2 a_n^2 \qquad \text{(vi)}$$

The total potential energy is obtained from equations (iv) and (vi) as

$$V = \frac{\pi^4 EI}{4L^3} \sum_1^\infty n^4 a_n^2 - W \sum_1^\infty a_n \sin \frac{n\pi c}{L} - \frac{\pi^2 P}{4L} \sum_1^\infty n^2 a_n^2 \qquad \text{(vii)}$$

From the principle of stationary potential energy we require that all the coefficients $a_n$ satisfy the condition

$$\frac{\partial V}{\partial a_n} = 0$$

thus from equation (vii)

$$\frac{\pi^4 EI}{4L^3} n^4 2 a_n - W \sin \frac{n\pi c}{L} - \frac{\pi^2 P}{4L} n^2 2 a_n = 0$$

or $\quad a_n = \dfrac{2WL^3}{\pi^4 EI} \dfrac{\sin n\pi c/L}{n^4(1 - P/n^2 P_E)}$

where $P_E (= \pi^2 EI/L^2)$ is the first Euler critical load (see *Essential Solid Mechanics*) for the beam-column in the absence of transverse load.

151

The deflexion under the load W is therefore given by

$$y(c) = \frac{2WL^3}{\pi^4 EI} \sum_{1}^{\infty} \frac{\sin^2 n\pi c/L}{n^4(1 - P/n^2 P_E)}$$

When $c = L/2$ and $P = 0$, the expression for the coefficients simplifies to that found in example 4.1. When $P \neq 0$, the effect of the axial load is to increase the deflexion due to the transverse load by the magnifying factor $(1 - P/n^2 P_E)$. As P approaches $P_E$, these deflexions become large enough to nullify the assumptions of the small deflexion theory.

*Example 4.7*

Determine the first critical load, $P_c$ for buckling of a pin-ended column of height L with a non-uniform circular cross-section whose second moment of area is given by

$$I = I_0 \left[1 + 4\left(\frac{I_c}{I_0} - 1\right)\left(\frac{x}{L}\right)\left(1 - \frac{x}{L}\right)\right]$$

where $I_c$ is the second moment of area at mid-height ($x = L/2$).

The deflected shape of the column under its lowest buckling load may be approximated by the first term of a sine series, thus a suitable displacement function would be

$$y(x) = a \sin \frac{\pi x}{L}$$

from which the strain energy due to bending is given by

$$U = \frac{\pi^4 E I_0 a^2}{2L^4} \int_0^L \left[1 + 4\left(\frac{I_c}{I_0} - 1\right)\left(\frac{x}{L}\right)\left(1 - \frac{x}{L}\right)\right] \sin^2 \frac{\pi x}{L} \, dx$$

Making the substitutions $x/L = \theta$ and $4[(I_c/I_0) - 1] = K$, we have

$$U = \frac{\pi^3 E I_0 a^2}{2L^3} \int_0^\pi \left(1 + \frac{K\theta}{\pi} - \frac{K\theta^2}{\pi^2}\right) \sin^2 \theta \, d\theta$$

From example 4.5, we have

$$\int_0^\pi \sin^2 \theta \, d\theta = \frac{\pi}{2}$$

$$\int_0^\pi \theta \sin^2 \theta \, d\theta = \frac{\pi^2}{4}$$

and
$$\int_0^\pi \theta^2 \sin^2 \theta \, d\theta = \frac{\pi}{12}(2\pi^2 - 3)$$

thus the strain energy finally becomes

$$U = \frac{\pi^4 EI_0 a^2}{4L^3}\left(1 + K \frac{(\pi^2 + 3)}{6\pi^2}\right) \tag{i}$$

The potential energy of the axial load P depends only on the form of the displacement function, thus

$$W_d = \frac{\pi^2 P a^2}{2L^2}\int_0^L \cos^2 \frac{\pi x}{L} \, dx = -\frac{\pi^2 P a^2}{4L} \tag{ii}$$

The total potential energy of the column is thus

$$V = \frac{\pi^4 EI_0 a^2}{4L^3}\left[1 + K \frac{(\pi^2 + 3)}{6\pi^2}\right] - \frac{\pi^2 P a^2}{4L} \tag{iii}$$

and for equilibrium

$$\frac{\partial V}{\partial a} = \frac{\pi^4 EI_0 a}{2L^3}\left[1 + K \frac{(\pi^2 + 3)}{6\pi^2}\right] - \frac{\pi^2 P a}{2L} = 0$$

or
$$a\left\{\frac{\pi^2 EI_0}{L^2}\left[1 + K \frac{(\pi^2 + 3)}{6\pi^2}\right] - P\right\} = 0 \tag{iv}$$

The two possible solutions of equation (iv) are

$$a = 0$$

or
$$P = \frac{\pi^2 EI_0}{L^2}\left[1 + K \frac{(\pi^2 + 3)}{6\pi^2}\right] = P_c$$

These two solutions are shown graphically in figure 4.9.

**Axial load, P**

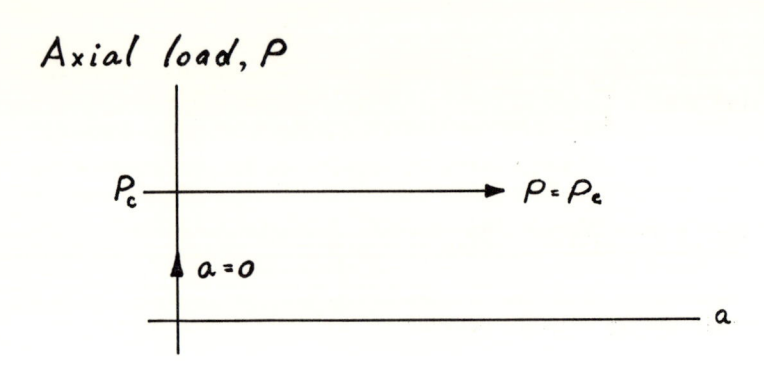

```
P_c ────────────────→ P = P_c
    ↑ a = 0
────┼──────────────────── a
    │
```

Figure 4.9

The point $a = 0$, $P = P_c$ represents a bifurcation, or branching, of the equilibrium state. When the axial load reaches the value $P_c$, the column is in a state of neutral equilibrium (see section 4.1) and the displacement a is arbitrary. Notice that by rewriting equation (iii) we have

$$V = \frac{\pi^2 a^2}{4L} \left\{ \left[ 1 + K \frac{(\pi^2 + 3)}{6\pi^2} \right] - P \right\}$$

thus V is a constant (zero) for both solutions of equation (iv).

Since the column has a circular cross-section we have

$$I_0 = \frac{\pi d^4}{64} \text{ and } I_c = \frac{\pi D^4}{64}$$

where d is the diameter at the ends and D is the diameter at mid-height.

Table 4.2 shows the first critical load for various values of the ratio D/d.

Table 4.2

| D/d | 1 | 1.1 | 1.2 | 1.3 | 1.4 | 1.5 |
|---|---|---|---|---|---|---|
| $P_c L^2/\pi^2 E I_0$ | 1.00 | 1.40 | 1.93 | 2.61 | 3.47 | 4.53 |

## 4.4 DESIGN EXAMPLE

A large travelling crane is mounted on a pair of rails each of which is supported on the ground through a series of transverse steel beams or cross-ties 10 m long. The greatest wheel load transmitted to the centre of a cross-tie is estimated to be 400 kN.  The ground is soft and reacts linearly to deformation.  Investigations have shown that the soil modulus is approximately 1 MN/m$^2$.

A suitable Grade 43 Universal Beam section is to be selected for the cross-ties.  In order to avoid excessive sinkage of the rail it is decided that the mid-span deflexion of the cross-tie relative to its ends should not exceed 1/360th of its span.

It may be assumed that the rail is bolted to the cross-tie.  Since the cross-tie will also sink into the ground when under load, sufficient lateral support is offered to eliminate the possibility of flexural-torsional buckling.  Thus, to ensure that the section may safely be designed elastically, the maximum stress due to bending is not to exceed 165 MN/m$^2$.

The problem simplifies to the design of a finite beam supported by an elastic foundation as shown in figure 4.10.

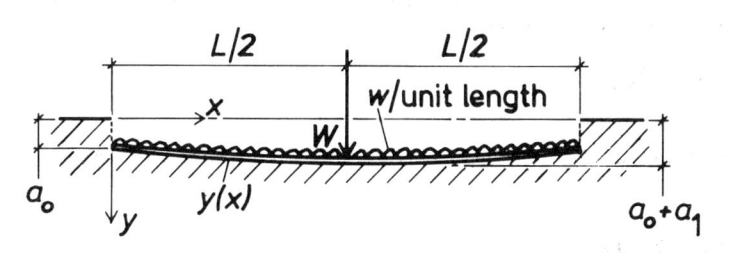

Figure 4.10

The wheel load transmitted by the rail is represented by the central concentrated load W.  The uniformly distributed load, w, arises from the self-weight of the beam.  Since the beam is relatively short it may be assumed that, under load, the ends sink below ground level by an amount $a_0$.  A suitable displacement function would therefore be

$$y(x) = a_0 + \sum_{1,3,5...}^{\infty} a_n \sin \frac{n\pi x}{L} \qquad \text{(i)}$$

The even sine terms are not required since they are asymmetric with respect to the beam centre-line.

From equation (i) we have

$$y'' = -\frac{\pi^2}{L^2} \sum_{1,3,5...}^{\infty} a_n n^2 \sin \frac{n\pi x}{L}$$

thus the strain energy stored in the beam is

$$U_b = \frac{\pi^4 EI}{2L^4} \int_0^L \left( \sum_{1,3,5...}^{\infty} a_n n^2 \sin^2 \frac{n\pi x}{L} \right)^2 dx$$

which, by virtue of the orthogonal properties of the sine function becomes

$$U_b = \frac{\pi^4 EI}{4L^3} \sum_{1,3,5...}^{\infty} n^4 a_n^2 \qquad (ii)$$

On a unit length of beam, the foundation exerts an upthrust q which is linearly dependent on the local deformation, y. If k is the foundation modulus, we have

$$q = ky$$

thus the strain energy stored in a small element of the foundation of length dx is given by

$$dU_f = \tfrac{1}{2}(q\,dx)y$$

The total strain energy stored in the foundation is thus

$$U_f = \frac{k}{2} \int_0^L y^2 \, dx$$

Substituting for y and integrating, we have

$$U_f = \frac{kL}{2} \left( a_0^2 + \frac{4a_0}{\pi} \sum_{1,3,5...}^{\infty} \frac{a_n}{n} + \tfrac{1}{2}\sum_{1,3,5...}^{\infty} a_n^2 \right) \qquad (iii)$$

The potential energy of the loads is given by

$$W_d = -w \int_0^L y\,dx - W\left( a_0 + \sum_{1,3,5...}^{\infty} a_n \sin \frac{n\pi}{2} \right)$$

or $$W_d = -wL\left( a_0 + \frac{2}{\pi} \sum_{1,3,5...}^{\infty} \frac{a_n}{n} \right) - W\left( a_0 + \sum_{1,3,5...}^{\infty} a_n \sin \frac{n\pi}{2} \right) \qquad (iv)$$

The total potential energy is therefore, from equations (ii), (iii) and (iv)

$$V = \frac{\pi^4 EI}{4L^3} \sum_{1,3,5...}^{\infty} n^4 a_n^2$$

$$+ \frac{kL}{2}\left(a_0^2 + \frac{4a_0}{\pi}\sum_{1,3,5...}^{\infty}\frac{a_n}{n} + \frac{1}{2}\sum_{1,3,5...}^{\infty}a_n^2\right)$$

$$- wL\left(a_0 + \frac{2}{\pi}\sum_{1,3,5...}^{\infty}\frac{a_n}{n}\right)$$

$$- W\left(a_0 + \sum_{1,3,5...}^{\infty}a_n \sin\frac{n\pi}{2}\right) \qquad (v)$$

For equilibrium, we require

$$\frac{\partial V}{\partial a_0} = \frac{\partial V}{\partial a_n} = 0$$

Hence

$$\frac{\partial V}{\partial a_0} = \frac{kL}{2}\left(2a_0 + \frac{4}{\pi}\sum_{1,3,5...}^{\infty}\frac{a_n}{n}\right) - (wL + W) = 0 \qquad (vi)$$

and $\quad \dfrac{\partial V}{\partial a_n} = \dfrac{\pi^4 EI}{4L^3} n^4 2a_n + \dfrac{kL}{2}\dfrac{4a_0}{\pi}\dfrac{1}{n} + \dfrac{kL}{4}2a_n$

$$- \frac{2wL}{\pi}\frac{1}{n} - W \sin\frac{n\pi}{2} = 0 \qquad (vii)$$

From equation (vi) we obtain

$$a_0 = \frac{(wL + W)}{kL} - \frac{2}{\pi}\sum_{1,3,5...}^{\infty}\frac{a_n}{n} \qquad (viii)$$

and from equation (vii)

$$a_n = \frac{\dfrac{2wL}{n\pi} + W \sin\dfrac{n\pi}{2} - \dfrac{2kL}{n\pi}a_0}{\dfrac{kL}{2}\left(1 + n^4\dfrac{\pi^4 EI}{kL^4}\right)}$$

After substituting for $a_0$ from equation (viii), this becomes

$$a_n = \frac{\dfrac{2W}{kL}\left(\sin\dfrac{n\pi}{2} - \dfrac{2}{n\pi}\right) + \dfrac{8}{n\pi^2}\sum_{1,3,5...}^{\infty}\dfrac{a_n}{n}}{\left(1 + n^4\dfrac{\pi^4 EI}{kL^4}\right)} \qquad (ix)$$

Evaluation of the coefficients $a_0$ and $a_n$ becomes tedious if more than one term of the sine series is included. In fact the series converges very rapidly and useful results may be obtained by assuming that $a_3$, $a_5$, etc., are all zero, hence

$$a_1 = \frac{\frac{2W}{kL}\left(1 - \frac{2}{\pi}\right)}{\left(1 - \frac{8}{\pi^2} + \frac{\pi^4 EI}{kL^4}\right)} \qquad (x)$$

and $\quad a_0 = \frac{w}{k} + \frac{W}{kL}\left[\dfrac{1 - \dfrac{4}{\pi} + \dfrac{\pi^4 EI}{kL^4}}{1 - \dfrac{8}{\pi^2} + \dfrac{\pi^4 EI}{kL^4}}\right] \qquad (xi)$

The only unknown quantities in the expressions for the two coefficients are w and I which depend on the choice of beam section. After inserting values for the known quantities, we have

$$a_1 = \frac{153.5}{1 + 0.103I \times 10^{-3}} \text{ mm} \qquad (xii)$$

and $\quad a_0 = w + 40 \left(\dfrac{0.103I \times 10^{-3} - 1.44}{0.103I \times 10^{-3} + 1}\right) \text{mm} \qquad (xiii)$

where the numerical value of w is the self-weight of the beam expressed in kN/m and the numerical value of I is the major second moment of area expressed in $cm^4$.

The maximum deflexion of the beam with respect to its end is $a_1$ and this is not to exceed L/360 thus

$$\frac{153.5}{1 + 0.103I \times 10^{-3}} < 27.8 \text{ mm}$$

hence

$I > 44027.2 \text{ cm}^4$

Referring to tables (see *Handbook on Structural Steelwork*, BCSA/CONSTRADO) we find that a 533 × 210 × 92 UB is the smallest section to satisfy this condition with a net second moment of area of 50 040 $cm^4$. Since w = 0.9 kN/m we now have from equations (xii) and (xiii) that

$a_0 = a_1 = 25$ mm

Since the second differential of the displacement function clearly does not have the necessary discontinuity at mid-span, a more accurate estimate of the bending moments in the beam is likely to be obtained by integration of the load diagram.

For one half of the beam, the total load p is given by

$$p = -(ka_0 - w) - ka_1 \sin \frac{\pi x}{L} \quad 0 < x < \frac{L}{2} \tag{xiv}$$

where downward forces are positive.

By integrating the expression for the load, we obtain the shear force

$$Q = \int p \, dx = -(ka_0 - w)x + \frac{ka_1}{\pi} L \cos \frac{\pi x}{L} + C$$

but $Q = 0$ when $x = 0$, thus $C = -ka_1/\pi L$ and

$$Q = -(ka_0 - w)x - \frac{ka_1}{\pi} L \left(1 - \cos \frac{\pi x}{L}\right), \quad 0 < x < \frac{L}{2} \tag{xv}$$

The bending moment is obtained by integration of the expression for the shear force given by equation (xv), hence

$$M = \int Q \, dx = -(ka_0 - w)\frac{x^2}{2} - \frac{ka_1}{\pi} L \left(x - \frac{L}{\pi} \sin \frac{\pi x}{L}\right) + D$$

but $M = 0$ when $x = 0$, thus $D = 0$ and

$$M = -(ka_0 - w)\frac{x^2}{2} - \frac{ka_1}{\pi^2} L^2 \left(\frac{\pi x}{L} - \sin \frac{\pi x}{L}\right), \quad 0 < x < \frac{L}{2} \tag{xvi}$$

The maximum bending moment occurs at $x = L/2$ and is given by

$$M_{max} = -\frac{L^2}{8}\left[(ka_0 - w) + \frac{4ka_1}{\pi^2}(x - 2)\right] \tag{xvii}$$

Substituting the values of $a_0$, $a_1$ and w corresponding to the 533 × 210 × 92 UB section, we find that $M_{max} = -466$ kN m. The elastic section modulus (Z) is 2072 cm$^{-4}$, thus the maximum stress due to bending is given by

$$\sigma_b = \pm \frac{M_{max}}{Z} = \pm 215.2 \text{ MN m}^{-2}$$

Unfortunately this stress is in excess of that permitted. It is therefore necessary to try a larger section. For a 533 × 210 × 122 UB we have

$I = 68719$ cm$^4$

$Z = 2794$ cm$^3$

and  $w = 1.2$ kN/m

hence, from equations (xii) and (xiii)

$a_1 = 19$ mm

and $a_0 = 29.1$ mm

From equation (xvii) the maximum bending moment is given by

$M_{max} = -458.6$ kN m

thus the maximum bending stress is

$$\sigma_b = \pm \frac{458.6 \times 10^{-3}}{2794 \times 10^{-6}} = \pm 164 \text{ MN/m}^2$$

therefore this section is satisfactory.

It is of interest to note that had the bending moment been calculated from $M = EIy''$, the maximum value (for the 533 × 210 × 122 UB section) would only have been -258 kN m.

There is, of course, an exact theory for beams on elastic foundations but in the case of finite beams of intermediate length (5 m<L<15 m for this problem) the solutions are tedious and the final results of great complexity.

As a check on the accuracy of the present solution, the exact values of $a_0$ and $a_1$ may be calculated (see *Advanced Strength of Materials*, Den Hartog, McGraw-Hill) for the 533 × 210 × 122 UB carrying a central concentrated load of 400 kN and supported by an elastic foundation of modulus 1 MN/m². The central deflexion ($a_0 + a_1$) is found to be 47 mm and the end deflexion ($a_0$) 28 mm. To correct for the effect of the beam's own weight we must add w/k (= 1.2 mm) to both of the above figures, thus

$a_1$ (exact) = 19 mm

and $a_0$ (exact) = 29.2 mm

These results are almost identical to those found by the energy method. We may therefore be reasonably confident that the values obtained for the maximum bending moment and the corresponding bending stress are also close to the true values.

A few further checks of the beam's adequacy remain (see *Handbook on Structural Steelwork*, BCSA/CONSTRADO). On the assumption that the web is unstiffened we require a 12 mm thick flange plate between the rail and the top flange of the beam in order to make full use of the direct bearing capacity of the web. There is no need to stiffen the web against buckling since with the addition of the flange plate the web buckling capacity is well in excess of 400 kN.

From equation (xv) the maximum shear force is found to be 200 kN. If we assume that all the shear force is carried by the web, the average shear stress is 31 N/mm$^2$. This is satisfactory since BS 449. The Use of Structural Steel in Building, allows an average shear stress of up to 100 N/mm$^2$ for Universal Beams of thickness less than 40 mm.

PROBLEMS

1. A simply supported beam of uniform cross-section and span L carries a concentrated vertical load at the left-hand quarter-span point. An additional flexible support of stiffness k is provided at mid-span. If the deflected shape of the beam may be approximated by the first term of a sine series, show that the force in the flexible support is

$$\frac{WL^3 k}{\pi^4 EI \left(1 + \frac{2kL^3}{\pi^4 EI}\right)} \sqrt{2}$$

2. A straight, uniform, elastic beam of flexural rigidity EI and length L is simply supported at its ends.

The deflected shape of the beam under a uniformly distributed load of w per unit length is correctly represented by the quartic

$$y(x) = a_0 + a_1 x + a_2 x^2 + a_3 x^3 + a_4 x^4$$

where y is the vertical deflexion at distance x from one end.

By minimising the total potential energy and noting that the quartic must satisfy the geometrical and statical boundary conditions, show that the maximum deflexion in the beam is given by

$$y_{max} = \frac{5wL^4}{384EI}$$

[Sussex]

3. A uniform beam of flexural rigidity EI and span L carries a uniformly distributed load of intensity w per unit length. The beam is rigidly built in at both ends in such a way that rotation and deflexion are fully restrained.

It is suggested that the deflexion of the beam may be approximated by the expression

$$y(x) = \frac{\Delta}{2}\left(1 - \cos\frac{2\pi x}{L}\right)$$

where the co-ordinate origin is taken at one end of the beam.

Investigate whether the above expression satisfies the necessary geometrical boundary conditions for the beam and if so, determine an approximate value for the central deflexion by minimising the total potential energy.

Show also that if the bending moment at mid-span is derived from the displacement function it will be approximately 19% in excess of the exact value. [$wL^4/4\pi^4EI$]
[Sussex]


4. A cantilever of depth d and length L tapers uniformly in width from B at the built in end to zero at the free end. Assuming that a vertical load W can be applied at the free end, determine the end deflexion Δ by minimising the total potential energy. It is suggested that the deflected shape of the beam be taken as

$$y(x) = \Delta \sin \frac{\pi x}{2L}$$

where the co-ordinate origin is at the free end.
[Sussex]         [$5.58WL^3/EBd^3$; the exact answer is $6WL^3/EBd^3$]

5. A uniform beam of span L is simply supported at its ends. By minimising the total potential energy, estimate the central deflexion under a non-uniform distributed load w given by the relationship

$$w = w_0 \sin \frac{\pi x}{L}$$

[Sussex]                                                    [$w_0L^4/\pi^4EI$]


6. A foundation pile of height L and flexural rigidity EI carrying axial load may be treated as a pin-ended column completely submerged in an elastic medium of lateral stiffness k per unit length. On the assumption that an infinite sine series is an appropriate displacement function, show that minimum buckling load is independent of L and the buckling mode and is given by $P_{cr} = 2\sqrt{(EIk)}$.

7. A pin-ended column carries an axial load and consists of three rigid bars of equal length h m connected end-to-end by elastic torsion springs of stiffness s kN m rad$^{-1}$. By minimising the total potential energy, show that there are two values of the critical load given by k/h and 3k/h kN.

8. A column in the form of a frustum of a solid circular cone of height h is built in at its base where the diameter is D and supported at the upper end (diameter d<<D) where an axial compressive force is applied. Assume a quartic polynomial as the displacement function and show, by minimising the total potential energy, that the critical load is $5\pi D^4E/48h^2$.

# INDEX

Approximate solutions 134
Axial loads 11, 149

Beams 81
Boundary conditions, curvature 134
    geometric 134

Castigliano 4
    first theorem (part I) 4
    first theorem (part II) 9
    second theorem 10
Compatibility equations 11
Compatibility method 5, 14
Complementary energy 1
    second theorem of 9
    stationary 7
Computer solutions 105
Crotti, Francesco 5
Curved members 86
    statically determinate 86
    statically indeterminate 92

Deflexions 50
Design examples 60, 114, 154
Dummy load 50

Energy due to axial forces 11
Energy due to bending 80
Energy theorems, auxiliary 9
    basic 2
    Castigliano's first theorem (part I) 4
    Castigliano's first theorem (part II) 9
    Castigliano's second theorem 10
    second theorem of complementary energy 9
Engesser, Friedrich 5
Equilibrium, neutral 132
    stable, unstable 6, 128
Equilibrium equations 11
Equilibrium method 5, 37

Flexible supports 32
Foundation modulus 155

Kirchhoff 6

Lack of fit 9, 22, 43

Moment distribution 105

Non-linear elasticity 17

Orthogonality    138

Plane frames, pin jointed    11
    rigid jointed    105
Potential energy    5, 128
    principle of stationary    6
Principle of least work    10
Principle of superposition    12

Rayliegh-Ritz method    134

Self-straining    9
Slope-deflexion equations    105
Statical indeterminacy    11
    conversion of systems    12
Strain energy    1
    direct use    50
    principle of stationary    10
Sway frames    109

Taylor's series    132
Temperature effects    27, 43, 61, 112